Sacred Flesh

The Science, Spirit, and Power of Intimacy

By: Deilen Michelle Villegas, Ph.D.

For permissions, inquiries, or speaking engagements, contact:
The Shamanic Goddess, LLC
Charlotte, North Carolina, US
www.DrDeilenMVillegas.com
Dr.Deilen.Villegas@Gmail.com

Book Title: Sacred Flesh: The Science, Spirit, and Power of Intimacy
Author: Dr. Deilen Michelle Villegas
Publisher: The Shamanic Goddess, LLC

Printed in the United States of America

ISBN: 978-1-969550-00-3
First Edition: 2026

Disclaimer

This book is intended for educational and informational purposes only. It is not a substitute for professional medical, psychological, or therapeutic advice, diagnosis, or treatment. The author is a Board-Certified Holistic Health Practitioner, Clinical Mental Health Counselor, Trauma Recovery Specialist, Clinical Sexologist, and Spiritual Consultant, and offers this material based on personal and professional expertise.

Readers are encouraged to consult with licensed professionals before applying any practices discussed in this book, especially if they have preexisting health conditions or are currently receiving psychiatric or medical care.

The author disclaims any liability or responsibility for any adverse effects resulting directly or indirectly from the use of information contained in this book.

Table Of Contents

- The Pain-Free Pleasure Map
- Sexual Safety Self-Assessment
- Safety Checklist for Sex During Menstruation
- Ritual for Reclaiming Menstural Wisdom
- Rewriting Your Sexual Beleifs- Self-Reflect Journal Prompts

Dedication

To the women who have forgotten they are divine, and to the men learning to love from their soul...

To the survivors, the sacred rebels, the seekers— who carry wounds in their bodies and wisdom in their bones.

To every soul who has ever been touched without consent, and to those reclaiming their bodies as temples, not battlegrounds.

This book is for the ones who are ready to unlearn shame, to remember pleasure as prayer, to embody love as liberation, and to make the bedroom a sanctuary again.

And above all, to the womb—the original altar— may we protect it, honor it, and birth more than just life from it. May we birth truth, healing, power, and sacred legacy.

With every word, this offering is yours.

— Dr. Deilen Michelle Villegas, Ph.D.

About the Author

Dr. Deilen Michelle Villegas, Ph.D.

Clinical Mental Health Practitioner. Holistic Trauma Specialist. Clinical Sexologist. Metaphysician. Spiritual Consultant. Sacred Storyteller.

Dr. Deilen Michelle Villegas is a Board-Certified Holistic Health and Wellness Practitioner, Clinical Mental Health Counselor, Trauma Recovery Specialist, Sexologist, Metaphysician and Spiritual Consultant whose work bridges science, spirituality, and ancestral wisdom. She is the visionary founder of **The Shamanic Goddess, LLC**, a transformational healing practice devoted to reclaiming wholeness in mind, body, and soul—especially within Black, Brown, and Indigenous communities.

With over **18 years of experience** in medicine—including Internal Medicine, Pediatrics, OB/GYN, Oncology, Behavioral Health and Urgent Care—Dr. Deilen brings a **deeply integrative approach** to healing. She holds multiple advanced degrees, including a Ph.D. in Organizational Leadership, an Ph.D. in Natural Medicine, M.A. in Clinical Mental Health Counseling, and an M.Sc. in Complementary & Integrative

Medicine. Her work is rooted in the belief that healing must be **culturally informed, spiritually aligned, trauma-responsive, and radically empowering.**

Her voice—bold yet soft, clinical yet poetic—invites readers into spaces of radical self-reflection, liberation, and sacred embodiment. Whether guiding clients through trauma healing, writing powerful literature on holistic wellness, or teaching the energetic anatomy of the womb and psyche, Dr. Deilen is devoted to one truth:

> *We do not heal to become someone new.*
>
> *We heal to return to who we've always been.*

She is also a devoted wife, mother of three, and fierce advocate for health equity, ancestral healing, and the liberation of feminine sensual power.

This book is one of many offerings in her growing body of work—a legacy of wisdom, sacred rebellion, and soul reclamation.

Connect with Dr. Deilen:

Website: www.drdeilenmvillegas.com

Instagram: @author_dr_deilen_m_villegas

Other Titles:

Reclaiming the Unspoken: The Journey Through Trauma, Truth, & Transformation

Rewired for Resilience: The Path to Neurosomatic Healing, Embodiment, and Transformation

Rooted In Wisdom: Holistic Herbal Medicine

For The Healing Communities

Awakened Science: Where Quantum Physics Meets the Divine

Encoded In The Stars: The Science of Destiny and the Blueprint of the Soul

Beyond Chemistry: Hormones, Healing & Human Connection in the Age of Disconnection

She Who Became: The Verses of a Wild Woman

Introduction

More Than Pleasure—Your Sacred Sexual Blueprint

Sex.
One of the most misunderstood, misused, and yet most powerful forces on this Earth.

We live in a world where sex is often reduced to a transaction, a moment of escape, or a performance. We've been taught to crave it, to fear it, to silence it, to weaponize it. But what few people teach us—what culture rarely honors—is that sex is not just an act. It is energy. It is communication. It is creation. It is a ritual.

As a Trauma Recovery Specialist, Clinical Sexologist, and Holistic Intimacy & Relationship Coach, I have sat across from clients whose relationships were shattered by unspoken wounds, by bodies that froze in moments meant for connection, by pleasure that felt more like punishment than peace. As a Spiritual Consultant, I've witnessed the other side—the unseen world of energy cords, soul ties, sacred unions, karmic loops, and rituals passed down through bloodlines. I've helped women reclaim their wombs as temples and helped men remember that their seed is a sacred offering, not a weapon of validation.

This book is born out of those experiences.
It is both a revelation and a reclamation.

You are holding a map that will take you deep into the neurological circuits of desire, into the somatic imprint of trauma, and into the ancient spiritual codes woven into our most intimate acts. You will read about oxytocin and orgasms, about Tantra and trauma, about pH balance and portals, about blood, bonds, and body memories. You will learn how sex can heal—and how it can harm. You will understand how the people we invite into our most sacred spaces can either unlock our destiny or delay it.

Because here's the truth: **Every sexual encounter is an energetic exchange.** And what you allow into your body... into your field... into your spirit... matters.

This book is not about shame. It's about sovereignty.
It's about moving from survival to sacredness.
From disconnection to deep divine union.

Whether you are healing from sexual trauma, rediscovering your sensual self, deepening the bond with your partner, or exploring the spiritual power of sacred sex—this book will meet you where you are and invite you to rise.

You are the portal.
You are the spell.
You are the sacred flesh.

Welcome to the journey of remembering.

— Dr. Deilen Michelle Villegas, Ph.D.

Chapter 1

The Neurobiology of Desire

Desire Begins in the Mind

Desire begins not in the body, but in the **neural corridors of the mind**—long before hands touch skin or lips meet lips.

Before the pulse quickens or pupils dilate, an intricate dance unfolds deep within the brain: a sacred, biochemical overture of longing. Every look, every scent, every brief brush of attention triggers an elegant flood of neurotransmitters and hormones. These are not just chemical reactions—they are the invisible authors of our attachment, arousal, craving, and remembrance.

Desire is not random. It is layered. It is intelligent.
It is encoded with memory, with meaning, with the stories your nervous system has never forgotten—even when your mind has.

This is the **neurobiology of desire**. And make no mistake: it is sacred.

The Brain on Love, Lust, and Longing

Neuroimaging studies have identified that romantic attraction and sexual desire engage multiple overlapping brain systems—primarily within the **limbic system**, which governs emotional processing, reward, and survival instincts. The **hypothalamus**, **amygdala**, and **nucleus accumbens** are activated during arousal and attraction, releasing a cascade of neurochemicals that shape how we interpret safety, intimacy, and pleasure.

Key brain structures involved in arousal:

- **Hypothalamus**: Regulates hormonal arousal and physical responses
- **Amygdala**: Processes emotion, including fear and attraction
- **Nucleus Accumbens**: Central to the reward system—lights up during pleasure and pursuit

Supporting Research:

- Fisher, H. E., Aron, A., & Brown, L. L. (2006). *Romantic love: A mammalian brain system for mate choice. Journal of Comparative Neurology, 493*(1), 58–62.
- Georgiadis, J. R., & Kringelbach, M. L. (2012). *The human sexual response cycle: Brain imaging evidence linking sex to other pleasures. Progress in Neurobiology, 98*(1), 49–81.

The Neurochemical Love Letter

When we experience sexual or emotional attraction, our brains release a powerful blend of neurochemicals that can create both euphoria and deep attachment. This isn't just chemistry—it's chemistry with purpose.

Primary Neurotransmitters and Hormones:

- **Dopamine**: Drives motivation, anticipation, and reward (the "craving" neurochemical)
- **Oxytocin**: Builds trust, empathy, and bonding (especially after orgasm or prolonged touch)
- **Serotonin**: Regulates mood, satisfaction, and impulse control

- **Endorphins**: Create pain relief and pleasure; support relaxation and bonding

- **Testosterone/Estrogen**: Drive libido and sexual receptivity

Together, these chemicals create a felt sense of intimacy and pleasure that can override logic, deepen emotional connection, and imprint powerful memories.

Further Study:

- Young, L. J., & Wang, Z. (2004). *The neurobiology of pair bonding. Nature Neuroscience, 7*(10), 1048–1054.

These neurochemicals don't just dictate who we're drawn to—they imprint the memory of that connection in our bodies and minds, for better or for worse. This is why sexual encounters—especially firsts, betrayals, and deep emotional connections—become etched into our nervous system. They form the *emotional blueprint* of our relational world.

Desire is not random.
It is a formula—a fusion of chemistry, personal history, trauma imprints, and conscious choice. And when we understand it, we gain the power to rewire it. To choose partners, not from our pain—but from our power.

Why Trauma Affects Desire

When individuals have experienced trauma—especially interpersonal or sexual trauma—the same neural systems involved in pleasure and bonding can become **hijacked**. The **amygdala**, for example, may become hyperactive and associate touch, intimacy, or arousal with fear rather than pleasure. The **prefrontal cortex** may go offline, and the body enters **fight, flight, or freeze**.

This helps explain why some trauma survivors feel shame or numbness around sex, or why they may find themselves stuck in patterns of hypersexuality or disconnection. The brain is not broken—it is adaptive. It learned how to survive.

Clinical Evidence:

- Lanius, R. A., et al. (2005). *Neural correlates of traumatic memories in PTSD. American Journal of Psychiatry, 162*(5), 970–976.

- van der Kolk, B. A. (2014). *The Body Keeps the Score.* Viking Press.

Rewiring Through Safe Connection

The good news? The brain can **rewire**. Neuroplasticity allows us to form new pathways when we are met with safety, slowness, and presence. When sexual or sensual experiences are **intentional, compassionate, and consent-based**, the nervous system begins to reassociate touch with connection—not threat.

Practices like:

- **Somatic therapy**: uses body awareness, breathwork and movement exercises to be more aware of bodily sensations and release stored emotions.
- **Trauma-informed tantra**: a mindful, body-centered approach that integrates ancient tantric practices wth modern trauma understanding, focusing on creating safety, releasing trapped energy (trauma) through breath, sound, touch (like de-armouring), and movement (shaking).
- **Mindful sensuality**: fully engaging your senses and being present with physical sensations, whether eating, touching, or during sex, to deepen pleasure and

connection.

- **EMDR and body-based psychotherapy**: psychotherapies that both target trauma's physical and emotional impact.

...can help the brain and body return to a regulated, receptive state. In this way, **pleasure becomes a portal for healing**, and connection becomes a corrective experience.

Desire Is Divine

Desire is not sinful. Desire is not shameful. Desire is sacred—it is the spark of creation, the whisper of the soul, the language of the body when it finally feels safe enough to speak.

To reclaim your sexuality is to **honor the holy intelligence** of your nervous system. To recognize that pleasure is not indulgence, but medicine. To know, without apology, that desire doesn't diminish your worth—it expresses it.

Trauma Bonding vs. True Connection

When the Body Confuses Chaos for Chemistry

Not all sparks are sacred. Not all butterflies are blessings. Sometimes, what we call "chemistry" is the nervous system replaying a **trauma script** it has never been taught to escape.

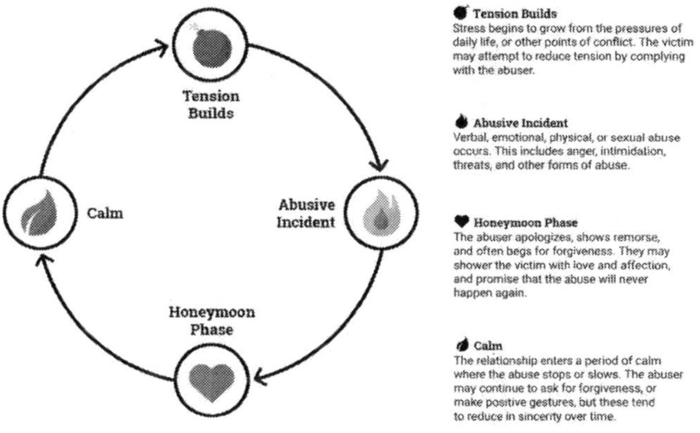

The body—especially when it has known neglect, abandonment, or volatility—may misinterpret chaos as passion. The familiar tension, the emotional rollercoasters, the push-pull dynamic—they don't always signal love. They can signal **dysregulation.**

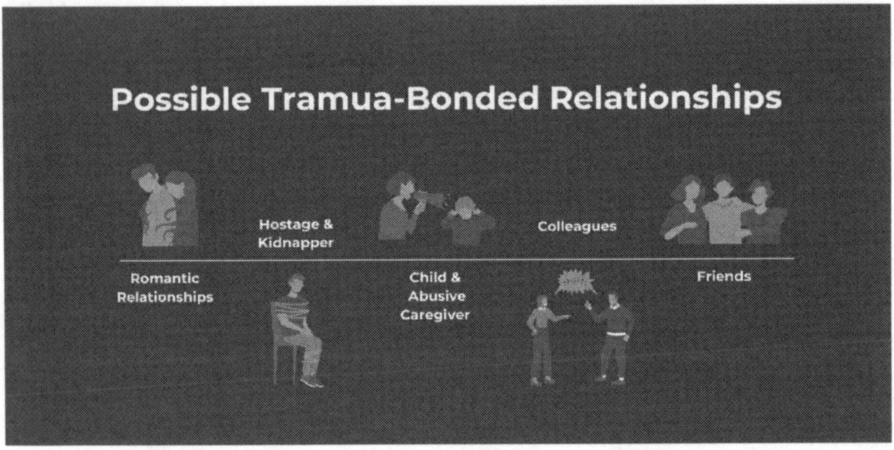

Trauma Bonding: Love Confused With Survival

Trauma bonding occurs when a person forms an intense emotional attachment to someone who is intermittently harmful. These bonds are not forged in safety—they are **wired in crisis**.

Biological Mechanism:

- **Adrenaline spikes** during moments of emotional chaos (fights, breakups, jealousy).
- This heightens **dopamine**, which the brain associates with reward.
- When reconciliation occurs (the "honeymoon phase"),

the **dopamine surge feels euphoric**—creating a **chemical high**.

- Meanwhile, **oxytocin** (the bonding hormone) still releases during intimacy, even with unsafe partners.

- Over time, this **reward-punishment cycle** becomes addictive. Like a gambler chasing a win, the brain begins to crave the **unpredictable emotional spikes**—not love, but biochemical survival.

Cortisol Hijack

High cortisol levels (chronic stress) impair the **prefrontal cortex**, which governs logic, decision-making, and impulse control. This is why people stay—even when they *know* the relationship is harmful. Their body is trying to survive a bond it doesn't feel safe leaving.

Clinical Evidence:

- Carnes, P. (1997). *Sexual addiction and trauma bonding. Counseling Today.*

- van der Kolk, B. A. (2015). *The Body Keeps the Score.*

- McCrory, E., De Brito, S. A., & Viding, E. (2010). *Research review: The neurobiology and genetics of maltreatment and adversity. Journal of Child Psychology and Psychiatry, 51*(10), 1079–1095.

How It Shows Up in Relationships:

- **Inconsistent affection** creates emotional addiction. You never know what version of your partner you'll get. But when they *do* show care, the reward feels magnified.

- **"Butterflies" are mistaken for love**—when in reality, they're your nervous system signaling anxiety or bracing for impact.

- **Jealousy, drama, and chaos feel exciting**, because the body has learned to associate stress with closeness.

- **You bond through survival**, not choice. And even when you're exhausted, you feel unable to leave.

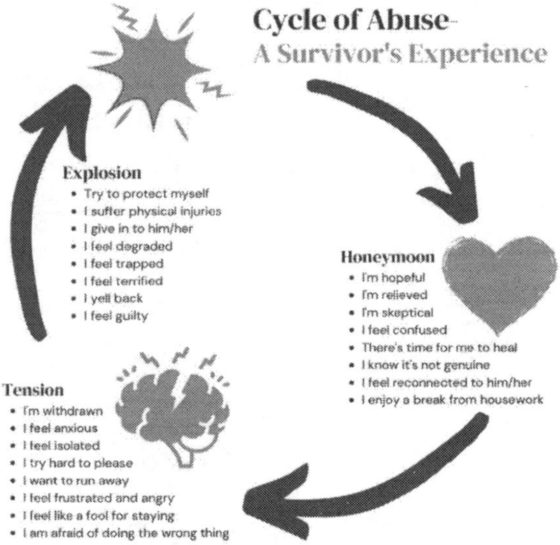

Healthy Attraction: Chemistry Without Chaos

Healthy chemistry is **steady**, not scarce. It builds over time, and is rooted in **co-regulation**, not cortisol.

What It Looks Like:

- **Oxytocin-driven bonding** (touch, trust, time) builds connection through safety.

- **Calm and arousal coexist**—you feel turned on *and*

emotionally grounded.

- There's **no performance** or pretending. You show up fully and are met in your wholeness.
- Your body doesn't brace. It relaxes.
- You stay, not because you're afraid to be alone—but because it feels good to be seen.

Research Spotlight:

- Coan, J. A., Schaefer, H. S., & Davidson, R. J. (2006). *Lending a hand: Social regulation of the neural response to threat. Psychological Science, 17*(12), 1032–1039.
  - This study found that simply **holding the hand of a loving partner** reduced neural responses to stress and threat.
- Acevedo, B. P., Aron, A., Fisher, H. E., & Brown, L. L. (2012). *Neural correlates of long-term intense romantic love. Social Cognitive and Affective Neuroscience, 7*(2), 145–159.
  - Evidence shows that **healthy romantic attachment** activates reward systems without the volatility seen in trauma bonds.

Ask Yourself:

- Am I chasing a feeling… or creating a foundation?
- Do I feel safe in my body when I'm with this person—or activated?
- Is this connection nourishing my growth or feeding my

fear of abandonment?

Understanding the difference is not just healing—

it's *liberation.*

From Chaos to Conscious Connection

Trauma bonding is not love—it's a **neurochemical survival strategy**. Liberation begins when you stop mistaking intensity for intimacy. True connection doesn't confuse you. It doesn't deplete you. It doesn't leave you waiting to be chosen.

True connection feels like:

- **Clarity** instead of confusion
- **Safety** instead of suspense
- **Warmth** instead of withdrawal
- **Consistency** instead of chaos

Healing the Pattern

Healing begins with **regulation**. If your nervous system has been trained to expect chaos, then love that feels calm may feel "boring" at first. But boring is not bad.
Peace is not a red flag.

As your body learns to feel safe again, you will stop craving the emotional spikes—and begin **craving depth, safety, and clarity**.

Healing Tools:

- Somatic therapy or EMDR
- Nervous system regulation practices (breathwork, grounding, cold exposure)

- Healthy relational templates (co-regulating with safe friends or partners)
- Education about trauma bonding and attachment styles

Liberation Is the New Love Language

You are not broken for craving intensity.
You are not foolish for falling into trauma bonds.
Your body was doing what it needed to survive.

But now it's time to **choose differently**.

You are allowed to outgrow adrenaline.
You are allowed to crave consistency.
You are allowed to say: "This isn't love. This is a pattern. And I'm ready to break it."

Trauma and Neurochemical Hijacking

When the Brain's Wiring Leads Us Back to Pain

The body doesn't forget. Even when the mind tries to move on, the nervous system remembers every betrayal, every abandonment, every violation.

This is the truth of trauma: it doesn't just haunt our memories —it reshapes our biology. And when it comes to intimacy, trauma often hijacks the very chemicals meant to create connection.

The Trauma Loop: A Nervous System on Repeat

In a healthy state, the brain's **prefrontal cortex** helps you make grounded, rational choices in intimacy. It evaluates safety, consent, desire, and long-term alignment.

But in the presence of trauma, especially unhealed childhood or relational trauma, that part of the brain gets overridden.

Instead, the **amygdala**—the brain's alarm system—takes control. Its job is survival, not connection. And it doesn't differentiate between the past and the present.

This is why:

- You may feel *unexplainably drawn* to emotionally unavailable or abusive partners
- You confuse *intensity* with intimacy
- You shut down during sex or feel numb when touched with tenderness
- You're addicted to the *chase*, but not the connection

These aren't character flaws. They're nervous system adaptations. The brain is replaying what's familiar—not because it's safe, but because it knows how to survive it.

Chemical Imbalances After Trauma

Trauma changes your neurochemical environment:

- **Low serotonin** can cause anxiety, depression, or obsessive attachment
- **High cortisol** (stress hormone) interferes with trust and pleasure
- **Dysregulated oxytocin** may impair your ability to bond or cause you to over-attach quickly
- **Adrenaline surges** may lead you to seek intensity instead of safety

Your body becomes wired to expect loss, betrayal, or abandonment—so even when safety shows up, it may feel boring or "off."

Sex as Survival: When Intimacy is a Trauma Response

Many trauma survivors use sex to:

- Feel a temporary sense of control or power
- Numb emotional pain or disassociate
- Earn love or approval
- Avoid being abandoned

In these moments, sex is no longer about pleasure or connection. It becomes a **survival strategy**—a way to feel something, anything, or to avoid feeling at all. And the cost? Deepened shame, misaligned soul ties, and a nervous system that continues to live in fear.

The Path Forward: Awareness as Alchemy

When you begin to recognize how trauma has shaped your sexual story, you reclaim the power to rewrite it.

You are not broken. You were wired for survival.
And now, you are safe enough to rewire for **sacred connection**.

From Survival to Connection: Rewiring Through Safety

Creating New Pathways for Love, Pleasure, and Presence

When intimacy becomes conscious, it becomes healing.

True safety isn't just the absence of conflict—
It's the presence of nervous system regulation. It's knowing that your "yes" is respected. That your "no" is honored. That your body does not have to shrink, perform, or disappear to feel loved.

When trauma has shaped our sexual experiences, we must begin the sacred work of **rewiring** the brain and body to receive connection without fear. This doesn't happen through logic alone. It happens through *felt safety*—over time, in the body, through practice.

Evidence-Based Healing Practices That Help Rewire Intimacy:

- **Somatic Therapy**
  Helps you become aware of how your body holds trauma, tension, and memory. It supports the release of stored pain while restoring a sense of grounded presence.

- **EMDR (Eye Movement Desensitization and Reprocessing)**
  A powerful trauma treatment that reprocesses painful memories and desensitizes triggers—so your past no longer controls your body's present reactions.

- **Mindfulness-Based Sexual Healing**
  Brings conscious awareness to touch, breath, and sensation. Helps you stay embodied and present during moments of pleasure or intimacy.

- **Tantric Breathing, Eye Gazing, and Sacred Rituals**
  Ancient practices that slow down time, increase oxytocin, and deepen connection—teaching the nervous system that pleasure can be slow, sacred, and safe.

When these practices are integrated with compassion and intention, the body learns something revolutionary:
That arousal can come from love—not fear.
That pleasure doesn't have to be earned through pain.

That connection isn't a performance—it's a presence.

You are not here to just survive touch.
You are here to be awakened by it.

The Spiritual Side of Brain Chemistry

Where Science Meets Sacred Energy

From a metaphysical lens, the chemistry of sex is not just biological—it's energetic. Every hormone, every neurotransmitter released during intimacy is not only altering your brain—it's activating your **energy body**. Your chakras —the subtle energy centers of your being—respond to emotional, physical, and spiritual states. During intimacy, these centers don't just flicker... they ignite.

- **Dopamine** awakens the **Solar Plexus Chakra**, your seat of willpower, confidence, and personal magnetism. This is the energy of pursuit, anticipation, and empowered desire.

- **Oxytocin**, the bonding hormone, floods the **Heart Chakra**, softening emotional barriers and creating space for trust, intimacy, and unconditional love.

- **Orgasm**, especially when fully embodied and energetically aligned, can burst through the **Crown Chakra**, opening a gateway to divine connection, bliss states, and even spiritual downloads.

When sex is approached as a sacred act—not a transaction—these chemicals become messengers of the soul.
The brain aligns with the spirit. The body becomes the temple.
The orgasm becomes the offering.

In this sacred state, sex becomes more than pleasure—it becomes **initiation**. A remembering. A portal. This is the spiritual side of neurochemistry: It's not just what's happening in the brain. It's what's awakening in your energy field.

Reflect & Reclaim

1. **When was the last time you felt truly safe during intimacy?**

2. **Have you confused adrenaline with love or connection in the past?**

3. **What stories do your desires tell about your nervous system's history?**

4. **What would it feel like to experience sex as a safe, slow, soul-deep experience?**

Your brain is not broken. It's brilliant.
Even in its adaptations—especially in its adaptations—it has been trying to keep you safe. Seeking love. Seeking connection. Seeking liberation.

When you begin to understand your chemistry, you reclaim your power. You reclaim your boundaries. You reclaim your *consent*—not just with others, but with yourself.

Because the most sacred, life-altering connection you'll ever experience...Is not with another person. It's the union between your mind, your body, and your spirit.

And when those three are in harmony—
you don't just experience intimacy.
You *become* it.

Chapter 2

Trauma, the Body, and Sexual Connection

When Touch Triggers Instead of Heals

You cannot heal what you continue to bypass. And you cannot experience sacred intimacy in a body that doesn't feel safe. For those who carry trauma—especially sexual, relational, or developmental—intimacy often feels more like a battlefield than a balm.

The mind may crave closeness, but the body flinches.
The heart longs to open, but the nervous system shuts down. Pleasure becomes panic. Desire turns into disconnection. This isn't dysfunction. This is protection.

The body is not broken—it's brilliant. It remembers what the mind tries to forget. It braces, flinches, freezes… not because it doesn't want love, but because it's still healing from the last time love hurt.

And until the body feels safe, no amount of mental readiness can force it to receive. Because intimacy isn't just about permission—it's about **presence**.

The Body Remembers What the Mind Tries to Forget

Trauma doesn't begin as a memory. It begins as a *sensation*—a shock to the system, a rupture in the body's ability to feel safe, seen, and in control.

And when the event has passed… the imprint remains.

Not always in words. But in breath that catches. In shoulders that curl inward. In a pelvis that tightens during touch. In eyes that dart or refuse to meet another's gaze.

This is not imagined. This is not weakness. This is **somatic memory**—the way your body holds on to what the mind cannot process.

The Science Behind Somatic Memory

When we experience trauma—especially events involving betrayal, violation, abandonment, or fear—the brain shifts into survival mode. The **amygdala** (the fear center) activates. The **prefrontal cortex** (the logic center) goes offline. And the **hippocampus** (the memory processor) struggles to store the experience properly.

What results is not just a cognitive memory, but a **body memory**. One that lives in the muscles. In the fascia. In the nervous system. In the womb. In the breath.

You might not *consciously remember* what happened,
but your body will—especially during moments of intimacy.

How Trauma Manifests During Sex or Touch:

- Sudden numbness or disassociation
- Inability to feel pleasure or orgasm
- Involuntary tears or anxiety when aroused
- Freezing during penetration or touch
- Feelings of guilt, shame, or emotional shutdown afterward
- Hyper-awareness of your partner's movements, sounds, or reactions

- Feeling the need to "perform" instead of connect

These aren't just emotional reactions. They are the **echoes of trauma** trying to protect you from being hurt again.

Your body is asking:

"Is it safe to be touched this way?"
"Will I lose control?"
"Do I have a voice in this moment?"
"Am I here because I choose to be—or because I feel I have to be?"

Your Body Is Not the Enemy—It's the Messenger

So many survivors are taught to "just get over it," to numb, to override, to perform. But you were never meant to override your body. You were meant to **listen** to it.
Because your body carries wisdom. And healing begins the moment we stop trying to silence it—and start honoring it.

You don't need to force yourself to be "ready" for intimacy. You need to create the safety required to receive it.
To know, in your bones, that your *no* will be honored...
That your boundaries matter...That your pleasure is not owed, but sacred.

The body will open again. But only when it knows it doesn't have to defend itself to survive.

Embodied Safety Ritual: Reclaiming Your Body's Yes and No

A gentle self-touch and visualization practice for reconnecting with your body and re-establishing safety.

You will need: A quiet space, 5–10 minutes of uninterrupted time, and your breath.

Step 1: Ground Into the Present

Sit or lie down in a comfortable position. Close your eyes. Take 3–5 deep, slow breaths.
With each exhale, imagine releasing tension.
With each inhale, invite your spirit to return to your body.

Place your hand over your heart or your lower belly—whichever feels most intuitively supportive. Feel your palm rise and fall with your breath.

Say (aloud or silently):

> *"This is my body. This is my space. I am safe to be here."*

Step 2: Connect With Your Edges

Now gently place one hand on your opposite shoulder or forearm.
Move slowly. Let your touch be soft, intentional.
Notice what it feels like to *choose* this touch. Not to perform, not to give—but to connect.

Ask yourself:

> *"Does this feel okay?"*
> *"Is there any part of me that wants more space, or a softer pace?"*

Listen to the response—without judgment.

Continue slowly exploring different areas: the collarbone, the thighs, the belly, the scalp.
Each time, ask your body:

> *"Do you feel safe?"*
> *"Do you want more, less, or something different?"*

Step 3: Affirm Your Sacred Boundaries

Place both hands back over your heart or womb space.

Affirm to yourself:

> *"My body speaks, and I listen."*
> *"I do not owe anyone my softness, my openness, or my yes."*
> *"I am allowed to move at the pace of safety."*

Finish with three grounding breaths. Wiggle your fingers, open your eyes, and return to your space with gratitude.

Journal Prompts For Reflection:

> "What did I notice about my body's yes and no today?"
> "What messages is my body trying to send me in moments of intimacy?"

Trauma Triggers in Intimacy

When Your Body Reacts Before You Can Explain Why

Trauma doesn't knock politely before entering the bedroom. It shows up suddenly. Loudly. Silently. It slips into the space between heartbeats and hesitation. It can turn a tender moment into a storm of confusion, fear, or shame. These responses are not irrational. They are **biological truths**—learned survival patterns embedded in the nervous system, preparing you for danger even when none is present.

Common Trauma Responses During Sex or Close Connection:

- **Freezing or Feeling Numb**
  Your body goes still. You mentally check out. You feel paralyzed. This is the dorsal vagal "shut down" state —your body's ancient way of self-preservation when escape feels impossible.

- **Sudden Tears or Panic**
  You may cry, tremble, or feel overwhelmed mid-

act, even if your partner is being kind. Your body is remembering something, even if your mind can't identify what.

- **Hyperawareness or Hypervigilance**
  You're constantly scanning for danger—reading your partner's face, body, tone. You're "in the act," but not *in your body*.

- **Anger or Emotional Overreaction**
  You may feel enraged by a simple touch, a comment, or a look. This is often your inner protector rising to defend your boundaries where they were once violated.

- **Inability to Orgasm or Feel Pleasure**
  Even when aroused, the body resists release. Orgasm requires surrender—and the nervous system cannot surrender where it does not feel safe.

- **Feeling Dirty, Ashamed, or Disgusted After Intimacy**
  You may consent to sex, but afterward feel regret, guilt, or emotional collapse. This can happen when a "yes" came from obligation, fear, or disconnection—not desire.

Why This Happens:

Trauma is *not* stored in your thoughts—it's stored in your **nervous system**. And during sex or emotional closeness, the body often interprets touch as a **threat**, not a gift—especially if it's familiar to a past wound. These are signs your **body is protecting you**, alerting you to where healing is still needed.

And until that trauma is met with compassion and regulation, the body will continue to react—even in the safest of beds, even in the arms of someone who loves you.

Important Truth:

You can want intimacy and still be afraid of it.
You can love someone and still feel triggered by their touch.
You can be aroused and still not be ready to be vulnerable.

These contradictions are not confusion. They are trauma's fingerprints—asking to be seen, heard, and healed.

Partner Communication Exercise

"What My Body Needs You to Know"

A trauma-informed tool for building trust, safety, and emotional connection in intimacy.

This exercise is designed to help partners open a dialogue around trauma, intimacy, and safety—without blame, pressure, or discomfort. It encourages vulnerability while honoring boundaries.

Create a safe container:

- Choose a quiet, private space with no distractions
- Sit facing one another with soft eye contact
- Set the intention: *"We are here to understand each other, not to fix or defend."*
- Hold hands or place a grounding object (crystal, cloth, or candle) between you to center the moment

Step 1: Take Turns Finishing These Sentences

(One person speaks while the other only listens. Then switch.)

"When I'm touched in a certain way, my body sometimes feels..."

"One thing I wish you knew when I get quiet or pull away is..."

"What helps me feel safe before or during intimacy is..."

"A gentle way you can check in with me is..."

"Something I'm still healing from is..."
"What I need most when I'm triggered is..."

Let this be a space for honesty—not performance.

Step 2: Create Your Shared Intimacy Agreements

Write down or discuss 2–3 agreements that support emotional and physical safety, such as:

- "We agree to slow down or pause if either of us feels overwhelmed."
- "We agree to check in verbally and energetically before moving forward."
- "We agree that intimacy is a choice—not an obligation."

These agreements create a co-regulated container for healing and connection.

Step 3: Closing Affirmation

Together, say:

> *"We are learning each other. We are listening to each other. And we are honoring the sacred space between us."*

Take a few grounding breaths and thank each other for showing up. Hold hands. Hug. Or sit in silence together.

Journal Prompt (Post-Exercise):

> "How did it feel to be heard? What surprised me about what I said—or what I learned about my partner?"

The Polyvagal System and Sexual Safety

How Your Nervous System Determines the Difference Between Love and Threat

You cannot have sacred sex in survival mode. You cannot access full-body pleasure if your nervous system is locked in protection. Your ability to connect, to trust, to feel present and embodied during intimacy depends on one essential system: **your vagus nerve**—the command center of your body's safety responses.

This is the heart of **Polyvagal Theory**, developed by Dr. Stephen Porges, which explains how your nervous system constantly scans for cues of safety or danger. It's called *neuroception*—and it's happening in the background, all the time, even during moments of connection.

The Three States of the Nervous System in Intimacy

Your body is always in one of three primary states. Understanding these states is essential for recognizing your capacity for intimacy, communication, and sexual connection.

1. **Ventral Vagal State – The Zone of Safety and Connection**

    - You feel calm, open, and grounded
    - Eye contact feels natural
    - Breath is steady and full
    - You're present in your body
    - Arousal feels exciting—not scary
    - Intimacy feels like a choice—not a demand

This is where sacred intimacy lives.
Here, you can be emotionally vulnerable, sexually open, and spiritually connected. This is the state where the heart and body say "yes" in harmony.

2. Sympathetic Arousal – Fight or Flight

- You feel anxious, restless, or irritated
- Your breath is fast or shallow
- You might feel hypersexual or completely disconnected
- You're on edge, trying to predict your partner's next move
- You may seek sex as a way to "release pressure" or avoid deeper emotion

In this state, sex becomes survival.
It may feel performative, aggressive, fast-paced, or emotionally numb. Your body is not relaxed—it's bracing.

3. Dorsal Vagal Shutdown – Freeze or Collapse

- You feel numb, distant, or emotionally unavailable
- You may go through the motions but feel "checked out"
- Touch may feel like pressure or even violation
- Orgasm may feel impossible, or followed by a wave of shame or sadness

This is the state of disconnection.
Here, the body has chosen shutdown to avoid overwhelm. Sex in this state often leads to confusion, regret, or emotional pain

—because the body is not consenting, even if the words say "yes."

Your Nervous System Always Comes First

You cannot "think" your way into intimacy. You must **feel safe** enough to surrender. This is why communication alone isn't enough.

Your body needs:

- **Consistency** (so it can stop bracing)
- **Slowness** (so it can catch up to your mind)
- **Choice** (so it can trust that it has power)
- **Repair** (so it can restore what was once ruptured)

Your Path Is Not Linear—It's Nervous System-Based

Some days you will feel open. Some days you will want to hide. This doesn't mean you're broken—it means your body is still learning what it's like to be in relationship *without danger*. The goal isn't constant arousal or orgasm. The goal is **attunement** —to yourself, and to your partner.

And that begins by honoring where you are—right now.

Healing Through the Body: Pleasure as Medicine

Reclaiming Sensation, Safety, and Self

After trauma, pleasure can feel distant. Not because we don't want it— but because our bodies no longer believe it is safe to receive.

So many survivors carry an invisible grief: A longing for touch, but a fear of what it might awaken. A hunger for intimacy, but a body that shuts down at the first sign of closeness. A memory of what was taken—and a question of whether it can ever return.

The answer is yes. **Yes, your pleasure can return.**
Yes, your body can become a source of joy—not pain.
But it won't come through force. It comes through *gentle reclamation.*

Pleasure Is Not a Performance—It's Presence

In a culture obsessed with orgasm, sex appeal, and perfection, we often forget that pleasure is not a goal. It is a *state of being.*

Pleasure begins with the smallest acts:

- A warm hand over your heart.
- The sun on your skin.
- A slow, deep breath that fills your belly.
- The smell of sacred oils or your favorite food.
- The sound of music that moves your spirit.

These moments tell the nervous system:

> *"It's safe to be here."*
> *"It's safe to feel."*
> *"It's safe to enjoy."*

And this is where healing begins.

Reconnecting to the Body Through Somatic Practice

Somatic healing means returning to the body—not to fix it, but to *feel it*. This process isn't linear. It's cyclical. Layered. Sacred.

Here are practices that can help restore your sense of safety and joy:

- **Body Scanning:** Gently tune in from head to toe. Ask, "What sensations are present?" No judgment. Just curiosity.

- **Consent-Based Self-Touch:** Let your hands explore your own body, not for arousal—but for connection. Let your body say yes. Let it say no.

- **Vagus Nerve Stimulation:** Humming, deep breathing, cold water on the face—these activate your body's calm state and help you regulate during or after intimacy.

- **Pleasure Mapping:** Slowly and safely discover what sensations, textures, temperatures, and types of touch feel good to *you.* Not to please another—but to rediscover *yourself.*

- **Sacred Bathing Rituals:** Use water, salt, herbs, or flower petals to cleanse your energy and reintroduce gentleness back into your skin and senses.

What Happens When Pleasure Becomes Medicine?

When pleasure is no longer attached to shame or expectation, it becomes healing. It tells the body a new story:
That connection is possible. That softness is allowed. That joy is not dangerous. That your body is worthy—not for what it does for others, but simply because it exists.

Pleasure is the language your soul uses to say:

> *"I am alive."*
> *"I am home in myself again."*

From Survival Sex to Sacred Connection

Rewriting the Story of Your Body, Worth, and Desire

For many trauma survivors, sex was never about pleasure. It was about protection. Control. Escape. Validation.
Survival. It became a currency for safety. A performance for approval. A mask to hide the pain.

But here's the truth no one told you:
You were never meant to survive sex. You were meant to be **transformed** by it. To be awakened by it. To be *witnessed* in your wholeness through it.

What Survival Sex Looks Like

- Saying yes when your body says no
- Using sex to avoid loneliness or abandonment
- Disassociating during intimacy but pretending everything is fine
- Performing pleasure instead of receiving it
- Feeling empty or ashamed afterward
- Believing that your worth is tied to being desired

These experiences are not shameful.
They are reflections of wounds—often passed down generationally, reinforced by patriarchy, colonization, religious trauma, or childhood abuse. But those wounds are not your identity. They are simply *where the healing begins.*

What Sacred Connection Feels Like

- Your body opens, slowly and by choice
- You are present, grounded, and attuned

- There is emotional safety—before, during, and after
- You communicate without fear of rejection or ridicule
- You experience mutual reverence, not performance
- Sex becomes a co-creative act, not a coping mechanism

This is sacred sex. Where the soul leads. Where the body softens. Where your truth is not hidden—but *honored.*

Reclaiming Your Sexual Power

Your healing is not about becoming perfect. It's about becoming **sovereign** in your body.

To say yes when it's a full-body yes.
To say no and feel no guilt.
To experience desire without shame.
To trust your body again.

To remember that your pleasure is not a sin—it is a spiritual right. And when you begin to live from this place of embodied truth...You no longer seek connection to be validated. You seek it to be *seen.* Because sacred connection is not just about who you're with. It's about who you are when you no longer abandon yourself to be loved.

You are not here to perform. You are not here to endure.
You are not here to be possessed, shamed, or silenced.

You are here to **embody joy.**
To **awaken your truth.**
To **honor the divine intelligence of your body.**

This is your reclamation. This is your return. This is the beginning of sacred connection.

Chapter 3

Arousal, Pleasure, and Communication Styles

Understanding How We Love, Respond, and Connect

There is no one-size-fits-all blueprint for desire. Arousal is not a switch—it's a language. And pleasure is not a performance—it's a deeply personal, sacred experience.

To reclaim our sexuality, we must first understand it. Not through the lens of porn, patriarchy, or programming—
But through *presence, curiosity, and compassion.* Because the more we understand how we respond to touch, attention, energy, and emotion… The more we can teach our partners how to love us **in our language**.

This is the path from confusion to clarity. From silence to sovereignty. From guessing… to guiding.

This is where pleasure begins to return home to the body.

Affirmation: I Give My Body Permission

Read aloud. Breathe it in. Let it become your truth.

I give my body permission to speak in whispers
and not be rushed.

I give my body permission to say no without guilt, to say yes without fear, to say not yet without question.

I give my body permission to want, to withhold, to wait—
to be wild, to be soft, to be still.

I release the need to perform. I release the shame inherited from silence. I release the belief that pleasure must be earned.

I return home to my own rhythm. I remember that my desires are divine. And I trust my body to guide me—
not through duty, But through deep, delicious truth.

What Is Arousal, Really?

Reclaiming the Truth About Desire

Arousal isn't about what turns someone else on. It's about what awakens **you.** Your rhythm. Your senses. Your yes. Arousal is not a destination. It's not a formula. It's not a performance.

Arousal is the body's way of saying:

> *"I feel safe."*
> *"I feel seen."*
> *"I want to open—not because I have to, but because I choose to."*

Arousal Is More Than Physical

Most people have been taught to associate arousal with one thing: the physical body.
Touch this, rub that, stimulate this nerve and voilà—desire.

But for many people—especially trauma survivors, women, neurodivergent individuals, and highly sensitive souls—**physical arousal is not the starting point.**

If emotional connection is missing...If trust is fractured...
If the energy feels off...Then no amount of physical stimulation will create authentic desire.

What Blocks Arousal?

- **Lack of emotional safety**
- **Unspoken resentment or relational disconnection**

- **Stress, overwhelm, or burnout**
- **Shame or religious conditioning around sex**
- **Nervous system dysregulation from unresolved trauma**
- **Feeling unseen, unappreciated, or objectified**

Your arousal is not resisting you. It's protecting you.
And when you start listening instead of forcing, your body will slowly begin to trust again.

The Invitation

Arousal is not about speed. It's about *attunement.* To your breath. To your boundaries. To your body's whisper before it ever becomes a scream. When you honor your unique arousal language, you become the authority of your own pleasure. And that's when sex stops being something that happens *to* you—and starts being something created *with* you.

The Five Types of Arousal

Your Desire Has a Language—Let's Learn It

Arousal isn't linear. It doesn't begin in the body for everyone. And it certainly doesn't look the same for every person—or in every season of life.

Just like we have different love languages, we also have different **arousal languages.** The more you understand your own, the more you can express it—and receive it.

Below are the five main types of arousal. You may relate to one, some, or all. None is more valid than the other. Your body holds the truth—this is simply a guide to help you listen more deeply.

1. Physical Arousal

You respond to touch, pressure, sensation, and stimulation.

- Erogenous zones (neck, inner thighs, nipples, ears) awaken desire
- Physical foreplay and skin-to-skin contact are key
- Massage, breathwork, or slow kissing may activate you
- You crave the sensory experience—the textures, the temperature, the movement

If physical arousal is primary for you, your body is the first door to your desire.

2. Emotional Arousal

You are most open to sex when you feel emotionally connected.

- Affection, vulnerability, and presence matter deeply
- You may not feel desire without intimacy, trust, and attunement
- Feeling cherished, appreciated, or affirmed can turn you on
- Hurt feelings or unspoken tension can shut down your body entirely

If emotional arousal leads for you, sex is less about the act—and more about the connection.

3. Mental Arousal

Your mind is your biggest turn-on.

- Deep conversation, shared dreams, or intellectual chemistry excite you
- Erotic stories, fantasy, imagination, or voice tone stimulate you
- Anticipation and buildup matter more than quick gratification
- You crave someone who can challenge or inspire you mentally

If mental arousal leads, intimacy begins in your thoughts, not your skin.

4. Spiritual Arousal

You crave energetic alignment and soul-level connection.

- Shared ritual, intention, prayer, or sacred touch arouses you
- You may experience sex as a portal or mystical union
- Eye gazing, breath synchronization, and heart connection are vital
- You often need a shared spiritual or energetic frequency to feel desire

If spiritual arousal leads, sex is a divine act—not just a physical one.

5. Energetic Arousal

You feel turned on through subtle energy, vibration, and presence.

- You sense attraction without touch—often just by someone's energy
- A glance, a breath, or being near someone can send energy through your body
- You're highly sensitive to your partner's mood, intention, and presence
- Overstimulation can shut you down—you need time and space to build

If energetic arousal leads, your body is tuned to frequency first, not friction.

You are not "too sensitive." You are *attuned.* Your arousal style is not strange. It is sacred. There is no wrong way to desire—only unexplored pathways to deepen it.

What's Your Arousal Style?

Discover how you uniquely experience desire and connection.

Instructions: For each of the following statements, check the ones that resonate with you. There are no right or wrong answers. Choose based on what feels true in your body right now—not what you think you *should* feel.

Physical Arousal

☐ I enjoy direct, physical touch as a way to connect.
☐ My body responds quickly to skin contact, pressure, or stimulation.
☐ I often need foreplay or sensual touch to feel turned on.
☐ I feel most connected during physical closeness.
☐ I enjoy exploring different textures, temperatures, or sensations.

Total for Physical: ____

Emotional Arousal

☐ I need to feel emotionally safe and connected before I can feel desire.
☐ If I feel unseen, hurt, or unappreciated, my arousal shuts down.
☐ Words of affirmation, affection, and emotional presence arouse me.
☐ I crave love, tenderness, or softness before physical touch.
☐ Intimacy without emotional connection feels empty to me.

Total for Emotional: ____

Mental Arousal

☐ I feel turned on by conversation, fantasy, or imagination.
☐ I enjoy teasing, anticipation, or slow build-up of desire.
☐ Deep conversation and mental stimulation open the door to intimacy.

☐ Erotic stories, intelligent humor, or a clever mind excite me.
☐ Sex without mental engagement can feel boring or disconnected.

Total for Mental: ____

Spiritual Arousal

☐ I crave energetic alignment or a soul connection with my partner.
☐ Sacred practices like breathwork, intention, or ritual deepen my intimacy.
☐ I view sex as a spiritual or energetic experience.
☐ I feel most alive when the emotional and spiritual realms are present.
☐ Union means more to me than climax—it's about sacred merging.

Total for Spiritual: ____

Energetic Arousal

☐ I can sense a partner's energy before they speak or touch me.
☐ Eye contact, presence, or energy can create arousal without physical touch.
☐ I am sensitive to overstimulation and need time to warm up.
☐ I often need space, pacing, and energetic trust to open.
☐ My body reacts more to intention than action.

Total for Energetic: ____

Results:

- Your highest score(s) reflect your dominant arousal style(s).

- Many people have a combination—it's normal to be multi-aroused!

- Use your results to open conversations, explore your

needs, and affirm your unique wiring.

Reflection Questions:

1. Which arousal style surprised you the most?
2. How have you been honoring (or ignoring) your unique arousal blueprint?
3. What do you want your partner(s) to understand about the way you connect?

Why Knowing Your Arousal Style Matters

The Path to Feeling Seen, Safe, and Fully Satisfied

Knowing how you experience desire is a revolutionary act of self-love. It breaks the cycle of confusion, shame, and silence. It stops the cycle of performance and starts the journey of **pleasure with purpose.**

Because when you don't know your arousal style…

- You may blame yourself for not "being in the mood"
- You may think something is wrong with you
- You may keep consenting to intimacy that doesn't actually feel good
- You may struggle to express what you need—so you stay silent
- You may begin to associate sex with stress, pressure, or emptiness

But when you *do* know your arousal style…

- You communicate your needs with clarity and confidence
- You stop forcing your body to match someone else's pace
- You release shame around how you love, connect, and awaken
- You build intimacy that's real, not rushed
- You create a connection that honors your body, your energy, and your heart

This is the Truth:

You don't owe anyone your arousal. You don't owe anyone your openness. You don't owe anyone your "yes."

Your job is not to mold yourself into someone else's idea of desire. Your job is to discover how *you* love. How *you* open. How *you* come alive. And from that knowing, you choose who is worthy of access.

You are not too complicated. You are complex. You are layered, textured, sacred, and whole.

When you know what awakens your soul, not just your body, you begin to experience sex, not as a transaction, but as a **transformation**. This is the root of sexual sovereignty. This is the beginning of sacred communication.This is where intimacy becomes alchemy.

Teach Your Partner Your Pleasure Language

How to Speak Your Needs Without Shame, Guilt, or Apology

Knowing your arousal style is powerful. But *teaching* it to someone? That's liberation. This isn't about demanding, dominating, or controlling the connection. It's about honoring your body's truth and inviting your partner into deeper intimacy—with *your consent* and *your clarity.*

Here's how to begin…

Step 1: Set the Energetic Container

- Choose a time outside of intimacy
- Eliminate distractions and drop into presence
- You might light a candle, play soft music, or sit face-to-face
- Say: *"I want to share something with you that's important to me—not because anything's wrong, but because I want to bring us closer."*

Step 2: Share Your Arousal Style(s)

Use "I" statements to explain what resonates with you and why. Be clear, but gentle. Empowering, not blaming.

Examples:

- *"I've learned that I'm emotionally aroused. When I feel connected through words or affection, my body opens more naturally."*
- *"Physical touch is a huge part of how I experience intimacy. Sometimes even a shoulder rub or back touch can build that*

connection for me."

- *"I love when we share deep thoughts or have playful conversations—it's not just fun, it actually helps me feel more turned on."*

- *"Spiritual intimacy is really important to me. Even eye contact or breathing together helps me feel aligned and connected."*

Step 3: Make It Collaborative

Turn this into an invitation—not a one-sided instruction.

Try asking:

- "What helps *you* feel connected?"

- "Is there anything I've done that made you feel deeply safe or seen?"

- "How do you like to be approached when you're not in the mood yet, but open to connection?"

- "Would you be open to exploring a ritual or practice together before we're intimate?"

Step 4: Co-Create Pleasure Agreements

These are simple, loving boundaries and rituals that support safety and connection.

Examples:

- "We check in with each other if either of us goes quiet or dissociates."

- "We don't rush to sex—we allow room for presence and exploration."

- "We always ask, *'Is this pace okay for you?'* before shifting intensity."
- "We remind each other: pleasure is co-created, not owed."

You are not asking for too much. You are asking to be known. And any relationship that's meant to hold your truth will make room for your voice.

Teaching someone how to love you is not selfish—it's sacred. And every time you do, you're not just honoring your body... You're giving your partner a chance to love you with reverence.

When Intimacy Misses the Mark: Mismatched Love Languages, Unmet Needs, and the Disconnect in Desire

Why Sex Isn't Working—and Why It's Not Always About the Act

Let's talk truthfully—intimacy struggles happen in even the most committed relationships.

So many couples are not lacking love...They are lacking **alignment**. They are giving from their own love language—but not *receiving* in the one that reaches them.
And when it comes to sex, this disconnect becomes even more painful.

When You're Speaking Different Love Languages

One of the most common issues I hear in sessions is this:

> *"I'm trying to connect, but I feel dismissed, ignored, or rejected."*

Often, what's happening is not neglect, but **mismatch.**

A partner whose love language is *physical touch* might initiate sex to feel close...While their partner, who needs *words of affirmation* or *acts of service*, feels unseen because their emotional needs aren't being met first.

So both leave the experience unfulfilled:

- One feels like a "selfish lover."
- The other feels used or emotionally bypassed.

This is why couples say:

> "We're having sex... but we're not connecting."
> "I give them what I think they want, but I'm not feeling desired myself."

You're going through the motions—but missing the meaning. Intimacy doesn't begin in the bedroom—it begins in the everyday exchanges that say, *"I see you. I value you. I choose you."* But what happens when those messages go unheard? When your love is speaking in French and theirs is in Morse code? The disconnect can feel like rejection... even when the love is still there.

The Breakdown Before the Bedroom

Many couples struggle with physical or sexual intimacy not because the desire is gone—but because emotional needs are unmet. When affection isn't communicated in a language the other person can recognize or receive, it creates emotional distance, frustration, and eventually, avoidance.

This isn't just a compatibility issue. It's a nervous system issue. Safety, connection, and erotic attunement require *resonance.* Without that, intimacy can feel like effort rather than ease.

What Are the 5 Love Languages?

Developed by Dr. Gary Chapman, the *Five Love Languages* framework outlines how individuals express and receive love. While not a diagnostic tool, this model provides a powerful lens for understanding relational misalignment and fostering deeper emotional intimacy.

1. Words of Affirmation

What it looks like:
Verbal expressions of love, praise, and appreciation. "I'm proud of you." "I love the way you handled that." "You're beautiful to me."
What it sounds like when missing:
"You never tell me how you feel about me anymore."
"I don't know if you still want me."

2. Acts of Service

What it looks like:
Doing thoughtful tasks that ease your partner's burden. Making dinner, running errands, or fixing something without being asked.
What it sounds like when missing:
"You say you love me, but you never *show* it."
"If you cared, you'd help out more."

3. Receiving Gifts

What it looks like:
Thoughtful tokens that show your partner you were thinking of them. Doesn't have to be expensive—just meaningful.
What it sounds like when missing:
"You never surprise me anymore."
"I don't feel cherished."

4. Quality Time

What it looks like:
Undivided attention. Shared experiences. Presence without distractions.
What it sounds like when missing:
"You're always on your phone."
"It's like we're in the same room, but miles apart."

5. Physical Touch

What it looks like:
Affectionate contact—holding hands, cuddling, kissing, and sexual connection.
What it sounds like when missing:
"I feel like you don't desire me."
"Why don't you touch me like you used to?"

Mismatched Languages = Missed Opportunities

Imagine trying to soothe a child in distress by offering a snack when they needed a hug. That's what happens in partnerships when we give love the way *we* understand it, instead of the way *they* receive it.

Love languages aren't static—they evolve based on trauma, attachment wounds, upbringing, and life stages. This means couples must stay *curious*, not just committed.

Just because someone shows love doesn't mean it's being felt.
And just because you're not "getting what you need" doesn't mean they don't care.

Intimacy Tip: Love Language Discovery Session

Make time for both partners to explore:

- **What makes you feel most loved?**
- **What tends to trigger rejection or disconnection for you?**
- **Has your love language shifted over time due to stress, trauma, or growth?**
- **Are you loving your partner in *your* language or theirs?**

Nervous System, Intimacy & Love Languages

Clinical studies show that emotional neglect or mismatch in relational needs can activate stress responses. When we don't feel seen or understood, our vagus nerve tone weakens, impacting co-regulation and arousal capacity.

Research Insight:

- Feldman, R. (2012). *Oxytocin and social affiliation in*

humans. Hormones and Behavior.

- Siegel, D. J. (2010). *The Neurobiology of "We": Relationship and the Developing Brain.*

Touch and attunement (especially when tailored to someone's love language) enhance vagal tone, oxytocin bonding, and safety in intimacy—both emotional and erotic.

Ask Yourself:

- What gestures *used to* make you feel loved that now feel absent?
- Have you clearly communicated your emotional needs—or assumed they "should know"?
- Are you open to learning your partner's love language without defensiveness?
- Are you using your partner's love language as a weapon or as a bridge?

This isn't just about romance. It's about restoration. When intimacy misses the mark, it's not a failure—it's a call to *get curious, not critical.*

How Men and Women Often Process Sex Differently

Of course, we all exist on a spectrum. But **biologically and neurologically**, men and women are often wired differently when it comes to arousal and connection.

***Men* (generally speaking):**

- Are more responsive to direct visual and physical stimulation

- Can become aroused and reach climax relatively quickly
- Often use sex as a way to access an emotional connection
- May not require a deep mental or emotional warm-up to enjoy the act

***Women* (generally speaking):**

- Require emotional and psychological arousal before physical engagement
- Orgasm is **80% mental**—meaning safety, presence, and emotional connection are critical
- If she's not mentally present, her body won't respond in the same way
- Needs time, foreplay, and trust to fully open

Let's Talk Science: Why Vaginal Penetration Isn't Enough

We need to dispel a dangerous myth:
Penetration alone is not the pinnacle of female pleasure—*and it was never meant to be.*

Most Women Do Not Orgasm from Penetration Alone

Clinical research has made this abundantly clear:

- **Only 18–25%** of women orgasm from vaginal intercourse without any clitoral stimulation.
- **Over 75%** of women require direct or indirect clitoral engagement to climax.

Research Insight:

- Lloyd, E. A. (2005). *The Case of the Female Orgasm: Bias in the Science of Evolution.*

- Herbenick, D., et al. (2017). "Women's orgasm: Body image and sexual function." *Journal of Sex & Marital Therapy.*

- Frederick, D. A., et al. (2017). "Differences in orgasm frequency among gay, lesbian, and heterosexual men and women." *Archives of Sexual Behavior.*

Why This Matters: It's Neurology, Not "Frigidity"

The clitoris contains **8,000 nerve endings**—more than any other organ in the human body. It exists purely for pleasure and extends deep internally, wrapping around the vaginal canal like a sacred halo of sensation.

The vaginal walls, in contrast, have fewer nerve endings—particularly in the upper third. They are more responsive to pressure, fullness, and emotional tone than to friction or speed.

This is why for many women:

- Thrusting doesn't equal arousal.

- Penetration without full-body readiness feels like intrusion.

- Faking pleasure becomes a coping mechanism for not being touched properly—or *at all.*

- Bigger does not always mean better. Women experience more pleasure with stimulation to the first-third of the vagina, where her (G-Spot) is located.

What True Readiness Sounds Like

Let's break down these poetic truths with science-backed respect:

"If she's not wet... she's not ready."
Lubrication is a physiological sign of arousal, but not a definitive one. Some women may produce less lubrication due to hormonal changes, medications, or trauma history. But generally, if the body isn't producing fluid, *slow down*. The nervous system needs more time to shift from sympathetic (guarded) to parasympathetic (receptive).

"If she's not opening... it's not time to enter."
Vaginal muscles contract reflexively when the body senses danger or discomfort. If penetration feels painful or blocked, it's a sign of muscular tension, unresolved trauma, or premature escalation. Consent isn't just a "yes"—it's a *softening*, an *opening*, a physiological yes.

"If she's never climaxed... she's not broken. She's likely untouched in the ways that matter most."
This is not failure. It's neglect—by lovers, by education, by society. True intimacy honors the entire erotic landscape: breath, skin, sound, slowness, and safety.

Anatomy of Attunement

True arousal isn't linear—it's layered.

- **Clitoral stimulation**: Includes external and internal components. May include circular motion, indirect touch, vibration, or oral engagement.

- **Emotional safety**: Felt through tone of voice, eye contact, and presence—not just physical touch.

- **Sensate focus**: A technique used in sex therapy to reconnect partners with non-goal-oriented intimacy and reduce performance anxiety.

- **Arousal pacing**: Women often need **20–40 minutes** of arousal before penetration feels fully welcome.

Therapeutic Reference:

- Basson, R. (2001). *Using a different model for female sexual response to address women's problematic low sexual desire.* Journal of Sex & Marital Therapy.

- Masters, W. H., & Johnson, V. E. (1966). *Human Sexual Response.*

Rewrite the Narrative

Too often, pleasure has been framed through a patriarchal lens —where speed, performance, and penetration are glorified, and deep feminine arousal is ignored or misunderstood.

But *sacred flesh* speaks a different truth:

> Pleasure is a journey, not a race. The clitoris is not a detour—it is the destination. If she hasn't bloomed, the soil may need tending—not the flower more pressure.

Arousal Invitation Practice

Invite your partner (or yourself) into the following:

- **Slow the pace** by 50%. Then slow it again.

- **Use your breath** to sync intention with presence.

- **Stay with her cues**, not your climax.
- **Build arousal with reverence**, not assumption.

When Sex Hurts Instead of Heals

Too often, penetration becomes the goal—rushed, assumed, expected—while presence, safety, and sensual attunement are overlooked.

But sex that ignores her readiness doesn't just miss the mark... It wounds what was meant to be worshipped.

What Happens When Intimacy Is Rushed or Misattuned:

- **Physical Pain**
  Without arousal, the vaginal tissues are less elastic, less lubricated, and more prone to:
    - **Tightness or burning**
    - **Micro-tears and inflammation**
    - **Pelvic floor spasms or vaginismus**
- **Emotional Shutdown**
  When the body is entered before the heart and mind feel safe:
    - Dissociation may occur
    - A woman may go silent, freeze, or "endure" the experience
    - She may question her own body or feel ashamed of her lack of pleasure

- **Long-Term Impacts**
  Repeated experiences of painful or unfulfilling sex can lead to:
    - **Sexual aversion**
    - **Low libido** rooted in trauma, not dysfunction
    - **Emotional withdrawal from the relationship**
    - **Unprocessed resentment and grief** stored in the womb space

Clinical Insight

Painful sex—also known as **dyspareunia**—is not uncommon. According to the American College of Obstetricians and Gynecologists (ACOG), nearly **3 in 4 women** will experience pain during intercourse at some point in their lives.

Sources:

- ACOG (2020). "Pain During Intercourse."
- Reissing, E. D., et al. (2004). "Prevalence, Characteristics, and Predictors of Sexual Pain Disorders in Women." *Sexual and Relationship Therapy*
- Brotto, L. A., et al. (2010). "A mindfulness-based group psychoeducational intervention targeting sexual arousal disorder in women." *Journal of Sexual Medicine*

Yet this pain is often **dismissed**, **normalized**, or **internalized**—with women being told to "relax more," "try harder," or "see a specialist," without partners being educated about the importance of emotional attunement and pacing.

The Importance of Warming Her Up—*Mentally, Emotionally, Energetically*

Arousal is not a switch—it's a spectrum. It doesn't start with touch—it starts with *tone*. With *trust*. With *time*.

Before the body opens, the nervous system must feel safe.

When this is honored:

- Muscles soften
- Breath deepens
- The vaginal canal lengthens and lubricates
- The mind shifts from guarded to receptive
- Orgasm becomes a *possibility*, not a performance

She Is Not "Too Sensitive"—She's Just Not Being Met Properly

Let's rewrite the narrative:

> If she tightens, it's not rejection—it's a request for deeper presence.
> If she pulls away, it's not withholding—it's a boundary saying: *"Not yet."*
> If she doesn't climax, it's not failure—it's feedback. She may need more tenderness, more curiosity, more connection.

Sex Is a Full-Body Conversation—Not Just a Physical Act

If intimacy is causing more harm than healing, something deeper is being bypassed.

Ask:

- Have we created emotional safety *before* we've created erotic stimulation?

- Are we prioritizing penetration… or presence?

- Do we listen to her body… or override it?

The Healing Alternative: Conscious, Attuned Sexuality

When done right, sex can *rewire trauma*, *restore worthiness*, and *rekindle joy*.

Healing sex honors consent, breath, slowness, and feedback. It builds arousal like a sacred fire—patient, intentional, divine. It doesn't just feel good—it feels **true**.

Healing the Disconnect Starts With Communication

Intimacy must be co-created—not assumed. We cannot expect our partners to be mind-readers. And we cannot take it personally if we've never been taught how to meet each other fully.

Here's how to begin rebuilding sacred connection:

- **Have regular, open conversations about pleasure, not just problems.**
  "What felt really good to you the last time?"
  "Was there anything you wished we'd done differently?"

- **Make space for guidance during intimacy.**
  Let your partner know it's *safe to speak up*—and *safe to be directed.* Learning each other's pleasure is not rejection —it's **reverence.**

- **Release the "one-size-fits-all" mindset.**
  Everyone's body is different. Every season is different. What turned you on at 25 may not work at 45. That's not

failure—it's evolution.

Intimacy isn't just about what you do. It's about how safe, free, and secure you feel while doing it. If you're not connecting, it doesn't mean you're failing. It means you're ready to deepen. To listen more. To give from understanding, not assumption. To receive in your full truth, not in silence.

Debunking the Myths: The Truth About Tightness, Virginity, and Vaginal Anatomy

Too often, women's worth and sexual "value" are reduced to unscientific, misogynistic, and harmful narratives—many of which stem from a lack of sexual education, patriarchal conditioning, and generational misinformation. Let's set the record straight.

Myth #1: "Tighter means she's a virgin. Looser means she's been with too many people."

False.

A woman's vaginal tone or "tightness" is *not* a reliable indicator of her sexual history, body count, or moral value. The vagina is not a one-size-fits-all canal with a permanent shape—it is a **dynamic, muscular, and responsive** structure designed to expand, contract, and adapt.

> What many men interpret as "tight" is often a sign that a woman is not aroused—or worse, bracing against discomfort.

Reality: Arousal Changes Everything

When a woman is **fully aroused**, her vaginal walls become lubricated, elongated, and *relaxed*—not tight. This allows for deeper penetration without pain, and signals that her **nervous system feels safe** enough to open and receive.

> If all your sexual partners "feel tight," it

may be less about their bodies—and more about the **lack of emotional connection, foreplay, and arousal** before penetration.

Myth #2: "Having babies stretches the vagina permanently."

Oversimplified.

While **childbirth** can temporarily stretch and weaken the pelvic floor muscles, this doesn't mean a woman's vagina becomes "loose" forever.

- With proper healing, **pelvic floor rehabilitation**, and time, most women's vaginal tone can return.
- In fact, many women become *more in tune* with their bodies post-childbirth—and learn how to engage their muscles consciously.

Reality: Pelvic Floor Muscles Are Like Any Other Muscle

They can be trained and strengthened with:

- Kegel exercises
- Pelvic physical therapy
- Breathwork and core activation

Just like doing squats builds leg strength, intentional pelvic floor work builds vaginal tone.

Medical Insight:

Vaginal tone is affected by:

- **Hormonal shifts** (e.g., postpartum, menopause)

- **Arousal levels**
- **Stress and trauma** (which can cause clenching or numbness)
- **Age and muscle elasticity**
- **Medical conditions** such as pelvic organ prolapse or connective tissue disorders

Sources:

- Bø, K. (2004). Pelvic floor muscle training is effective in treatment of female stress urinary incontinence. *International Urogynecology Journal*
- Masters, W. H., & Johnson, V. E. (1966). *Human Sexual Response*
- Kingsberg, S. A. (2002). The impact of aging on sexual function in women and their partners. *Archives of Sexual Behavior*

Reframing the Conversation: From Objectification to Understanding

Instead of focusing on "tightness" or outdated notions of virginity, let's shift toward more empowering, body-literate conversations:

Ask:

- Is she **emotionally ready** and *enthusiastically consenting*?
- Have we created **foreplay that engages her mind, heart,**

and body?

- Is her **pleasure prioritized**, not just her participation?
- Are we measuring intimacy by depth of connection—not by constriction?

Harmful Commentary to Retire:

Statements like:

- “She’s too loose”
- “She felt like she’s had a lot of partners”
- “She didn’t feel tight, so it wasn’t good”

...are rooted in ignorance, disrespect, and internalized misogyny. These ideas not only shame women—they rob men of the chance to experience true intimacy and connection.

The Sacred Truth:

- A woman’s worth is not between her legs—it’s in her **sovereignty**.
- Arousal is not performance—it’s a **physiological response to safety and trust**.
- True lovers learn their partner’s body—not by judgment, but by presence.

A relaxed vagina isn’t “loose”—it’s **ready**. And readiness is sacred.

Couples Intimacy Check-In Worksheet

Reconnecting Through Reflection, Vulnerability & Intentional Love

> "When you don't talk, your body will. When you don't feel heard, your intimacy will reflect it."

Use this check-in to explore emotional needs, intimacy preferences, and pleasure desires. Fill it out individually, then come together to share.

Section 1: Emotional Connection

1. Lately, I've felt most connected to you when...

☐ You touched me lovingly
☐ You affirmed or encouraged me
☐ We laughed or played together
☐ We spent intentional time without distractions
☐ You supported me during a hard time
☐ Other: ______________________

2. One thing I need more of emotionally is:

____

____

____

3. One thing I've been hesitant to share with you is:

____

____

____

Section 2: Sexual Intimacy & Pleasure

4. When we're intimate, I feel most aroused by...

☐ Emotional closeness
☐ Foreplay and slow build-up

☐ Sensual or erotic touch
☐ Eye contact and deep breathing
☐ Clitoral stimulation / oral sex
☐ Talking about fantasies or desires
☐ Other: ____________________

5. Something I'd like to explore more together is:

____

6. Something I'm afraid to ask for but secretly desire is:

____

Section 3: Communication & Growth

7. The way I receive love best right now is through...
☐ Words of affirmation
☐ Acts of service
☐ Physical touch
☐ Quality time
☐ Gifts / thoughtful gestures
☐ Presence and attentiveness

8. The way I express love most naturally is through...
☐ Words of affirmation
☐ Acts of service
☐ Physical touch
☐ Quality time
☐ Gifts / thoughtful gestures
☐ Other: ____________________

9. What I need from you in moments of conflict or misunderstanding:

____

Section 4: Sacred Agreements

Together, write 2–3 loving commitments to support your intimacy moving forward:

1. *We agree to* ______________________________

2. *We will make space for* ______________________________

3. *We will honor each other's pleasure by* ______________

Closing Ritual

Sit facing each other. Place your hands over your heart or hold hands. Make eye contact. Repeat aloud or in your heart:

> *"I see you. I choose you. I am learning how to love you more deeply every day."*

Take 3 deep breaths together to anchor the moment.

Pleasure vs. Performance

Reclaiming Intimacy from Expectation, Ego, and Shame

We can be honest—so many people are having sex,
but very few are actually experiencing **pleasure.**

That's because we've been conditioned to perform. To fake it. To please without receiving. To "climax" on cue.
To look sexy instead of *feel* connected. Sex has become a stage for many, rather than a sanctuary.

But real pleasure doesn't live in performance. It lives in **presence.** It lives in safety, slowness, self-trust, and surrender.

Signs You're Stuck in Performance Mode:

- You feel pressure to "do it right" or "keep your partner satisfied"
- You fake orgasms or reactions to avoid awkwardness or disappointment
- You worry more about how you look than how you feel
- You check out mentally and go through the motions
- You ignore pain or discomfort to avoid conflict or guilt
- You initiate or accept sex out of obligation, not desire

Performance is often a trauma response. It's survival masquerading as seduction. It says, *"If I give you what you want, maybe you'll stay."*

But survival is not the same as intimacy. And pleasing is not the same as pleasure.

Signs You're Rooted in True Pleasure:

- You feel safe to go slow—or to stop
- You're connected to your breath, your body, your voice
- You express your desires without guilt
- You stay present with your sensations
- You feel nourished afterward, not depleted or confused
- You know that your orgasm is not the measure of your worth

The Nervous System Knows the Difference

Even if your face smiles, even if your mouth says "yes"—
your body knows when it's pretending. Your breath knows when it's bracing.
Your muscles know when they're being used versus cherished.
True pleasure requires presence. Not just of the body, but of the **soul**.

Important Truth:

> *If you're not enjoying it—it's not intimacy, it's performance. If you're disconnected from your own desire—it's not empowerment, it's endurance.*

And if you've been there—you're not alone. Many people have been conditioned to prioritize someone else's pleasure, while never exploring their own. But it's not too late to return to yourself.

Your pleasure is not selfish. It's sacred. And you don't have to earn it through perfection, silence, or sacrifice. You are allowed to be slow. You are allowed to need more.

You are allowed to receive without guilt. Because you were not born to perform. You were born to *feel.*

The Power of Pleasure-Centered Communication

Speak Your Needs. Honor Your Body. Deepen Your Connection.

You deserve to be known. Not just tolerated. Not just touched. But **understood**—in the language your body speaks.

You deserve to be touched in the way your body *responds to* —not the way someone assumes it should be touched. You deserve to speak your needs without shrinking. Without guilt. Without apologizing for the depth of your desire. You deserve to stop explaining yourself as if your pleasure is a burden. But this level of intimacy requires communication that is: **Clear. Compassionate. Courageous.**

Real intimacy isn't just built in the bedroom—

It's built into the conversations that happen *before* you ever undress. When you use your voice, you're not rejecting your partner. You're **inviting** them into deeper access. And when they listen—they're not being controlled.
They're being taught how to worship what's sacred.

Phrases to Use for Communicating Desires and Boundaries

Because Your Voice Is Sacred—and So Is Your Body

Speaking your truth during intimacy doesn't make you high-maintenance. It makes you **honest.** And honesty is the bedrock of sacred connection. Whether you're guiding a partner, setting a boundary, or expressing a new desire, use these phrases as conversation starters, permission slips, and sacred declarations.

Phrases for Expressing Desire

> "I really enjoy it when you..."

"Right now, I'd love for you to..."
"Can we go slower? It helps me stay present."
"I'm feeling really connected—can we explore more?"
"I've been thinking about something I'd love to try with you..."
"When you do ___, my body really responds to it."
"Would you be open to trying ___ together?"
"Can we breathe together before we start? It helps me drop in."
"I want to feel closer—can we spend more time touching or cuddling first?"

Phrases for Setting Boundaries (Without Shame)

"I'm not ready for that right now, but I'd love to stay connected in other ways."
"That doesn't feel good to me—can we try something else?"
"I'm feeling a little disconnected. Can we pause and check in?"
"Please slow down. I need a moment to feel safe in my body."
"I love being with you, and I want to honor what feels right for me."
"If I go quiet, that might mean I'm leaving my body. Can you help me stay grounded?"
"I'm open to touching, but not penetration right now."
"I need more emotional connection before moving forward physically."
"Let's pause. I want to be fully present, and I'm not there yet."

Phrases for Checking In With Your Partner

"How are you feeling in your body right now?"

"Is there anything you want more or less of?"
"Do you feel safe with me in this moment?"
"Is there anything I can do to help you feel more connected?"
"Are you okay with the pace we're moving at?"
"What would help you feel more relaxed or open?"
"Do you want to guide me with your words or hands?"
"Is there anything you've been wanting to explore but haven't said yet?"

Speaking during intimacy isn't a disruption—it's a deepening. Every word is a thread that weaves safety, trust, and deeper pleasure into your connection.

Your body is sacred. Your voice is holy. Your truth is a love language.

Laughter Is Intimacy Too

The Sacred Power of Lightness in the Bedroom

Here's the truth no one talks about enough:
Sex doesn't always go as planned.

Sometimes a sound slips out. Sometimes a leg cramps. Sometimes you miss the rhythm. Sometimes you just can't stop giggling.

And that's okay. It doesn't mean the moment is ruined. It doesn't mean the intimacy is lost. In fact, **those moments are intimacy.**

When You Can Laugh Together, You're Safe Together

Laughter isn't rejection—it's **release**.
It's the body saying, *"I trust you enough to be real."*

When you and your partner can laugh *with* each other, not *at* each other, you're cultivating:

- Emotional security
- Nervous system regulation
- Deeper relaxation and comfort
- A foundation of *friendship* within your erotic connection

Try Saying:

> "Well… that wasn't in the script—but I love that we can laugh about it."
> "We're human, not porn stars—and honestly, this makes me feel closer to you."
> "If we can laugh together during sex, we're definitely doing something right."

Your intimacy doesn't have to be flawless to be **beautiful**. It doesn't have to be serious to be **sacred**. The most powerful sex happens when you can be soft, silly, and real—without fear of judgment.

Because laughter…is foreplay, is connection, is liberation.

Chapter 4

Energy Transfer, Absorption & Sexual Hygiene

When Sex Leaves More Than Just a Memory

Every sexual experience leaves a fingerprint on the soul. Not just a memory in the brain, but an **energetic imprint** in the body. We are not just flesh. We are frequency. And sex, at its core, is an energetic exchange— a merging of aura, intention, vibration, and spirit.

When we engage in sex, we are not just exchanging fluids or pleasure. We are exchanging **energy**—stories, wounds, beliefs, attachments, ancestral patterns, even entities or emotional imprints. And the more conscious we become of this truth, the more intentional we become with *who* we allow into our sacred space.

What Is Sexual Energy Transfer (Exchange) S.E.X.?

Sexual energy transfer is what happens when two (or more) energetic bodies merge during intimacy. This goes beyond touch—this is **energetic blending**.

What gets exchanged?

- **Emotions** (love, shame, grief, joy)
- **Intentions** (dominance, devotion, manipulation, healing)
- **Spiritual residue** (entities, attachments, generational pain or power)

- **Belief systems** (subconscious programming around worth, control, love)

- **Physical frequency** (fatigue, vitality, illness, or creative energy)

Your body is like a sponge. It absorbs what is not yours —especially when your boundaries are weakened by arousal, trauma, or emotional enmeshment.

Signs You've Absorbed Someone's Energy During Sex

- You feel heavy or drained afterward

- You suddenly feel emotions that aren't yours (sadness, anger, anxiety)

- You're mentally replaying the experience or feel "hooked" to the person

- You begin to carry their energetic burdens—dreams, habits, or even illnesses

- You have trouble focusing, feel disconnected from your purpose or creativity

- You keep energetically returning to the person, even if it's over

This isn't obsession. It's **residue**. And if not cleared, it can disrupt your mind, body, and spirit for weeks, months, even years.

The Science of Vibration—How Arousal Shifts Your Frequency

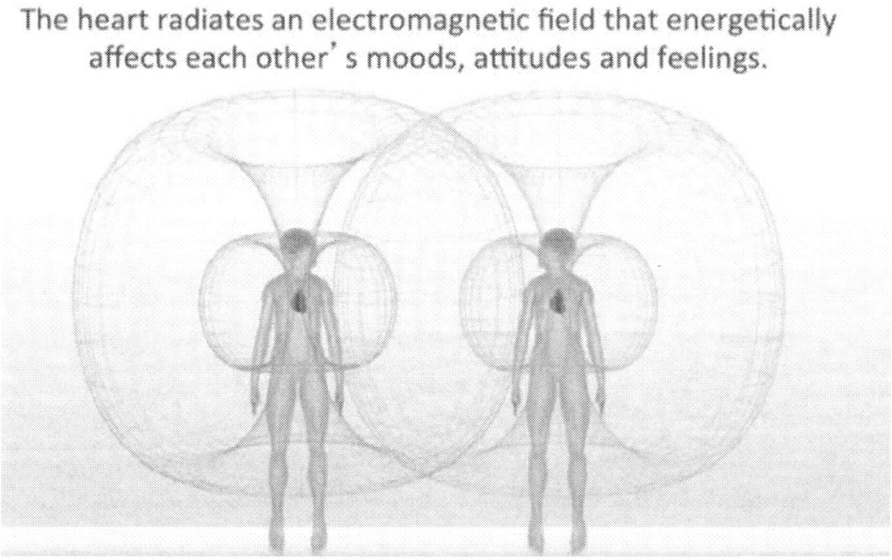

Your body is electric. Your arousal is energetic.

Every human being emits a **vibrational frequency**—an energetic signature made up of electrical impulses from your nervous system, electromagnetic fields from your heart, and the biophotons emitted by your cells.

According to biophysics and quantum biology:

- Your **heart** emits the largest measurable electromagnetic field in the body.

- Your **brainwaves** change during arousal—moving from beta (thinking) to alpha/theta (relaxed, meditative states).

- Your **nervous system** shifts from sympathetic (fight/flight) to parasympathetic (rest/pleasure) when you're emotionally and physically safe.

- At the moment of orgasm, there is a **neurochemical and vibrational spike** that can register as a *brief altered* *state of consciousness*—heightened awareness, euphoria, or even visions.

During Sexual Arousal:

- Your **energetic field expands**, increasing sensitivity to external energy

- Your **chakras activate**, especially the sacral (creativity/pleasure), heart (love/bonding), and crown (spiritual awareness)

- Your body emits more **electrical charge** and **heat**, creating a literal "charge" between you and your partner

- If emotional or spiritual alignment is absent, this charge can become **chaotic**, leading to energetic confusion, depletion, or spiritual entanglement

The Bottom Line:

Your energy field becomes more **open, receptive, and impressionable** during intimacy. This is why sex can either **recharge** you... or **drain** you. It all depends on intention, presence, emotional safety, and alignment.

Journal Reflection: Energetic Boundaries & Past Partners

Calling Your Energy Back. Reclaiming Your Sacred Space.

Sex is not just physical—it's a portal. And sometimes, even when a relationship ends, the **energy exchange lingers.**

This journal reflection is your invitation to explore what you've absorbed, what you're still carrying, and what you're ready to release.

Set aside quiet time, light a candle, and breathe deeply before writing. Let your truth speak without judgment.

Reflect On:

1. Which past partner(s) do I still feel energetically connected to—emotionally, mentally, or physically? Why?

2. After intimacy with someone, what do I usually feel—drained, nourished, clear, confused?

3. Were there times I gave my body but not my energy willingly? How did that impact me afterward?

4. Have I ever taken on someone else's pain, sadness, anger, or chaos after sex?

5. What patterns or wounds am I ready to stop recycling through sexual connection?

6. What does energetic safety mean to me now? What do I need to feel protected, present, and sovereign during sex?

Ritual Closing (Optional):

Write this or say aloud:

> *"I call my energy back from every person, place, or moment that is no longer aligned with my highest truth. I bless what was. I release what no longer serves. I am whole. I am sovereign. I am sacred."*

Sexual Energy Cord-Cutting Ritual

Reclaiming Your Power. Releasing What No Longer Belongs.

> *"Every time you return your energy to yourself, you reclaim your power. You are not just letting them go—You are letting* ***you*** *come home."*

This ritual is for releasing residual energetic ties, imprints, and entanglements created through sexual or intimate connection. It's not about denying your past—it's about clearing your field so you can fully inhabit your present.

You can do this alone or with the guidance of a trusted healer or practitioner. Perform this ritual when you feel ready—not rushed.

You Will Need:

- A white candle (purity, clarity)
- A black candle (release, transformation)
- A bowl of water or salt (cleansing)
- Journal or paper
- Optional: sacred herbs (sage, mugwort, palo santo), rose petals, or a crystal (black obsidian, selenite, or rose quartz)

Step 1: Prepare the Sacred Space

- Cleanse the space using your chosen herb or incense.
- Light both candles.
- Sit comfortably with your journal, eyes closed, and take

7 deep breaths.

- Invite your spirit team, ancestors, or highest self to guide and protect you.

Say aloud:

> *"I enter this space with reverence. I am ready to release all that no longer serves my body, energy, and spirit."*

Step 2: Name Who or What You're Releasing

In your journal, write the names or initials of any person(s) with whom you feel a lingering energetic tie—particularly those formed through sexual or intimate exchange.

Next to each, write a statement like:

- *"I release the emotional weight I've carried from this connection."*

- *"I forgive myself for what I gave away when I didn't feel fully safe."*

- *"I release the belief that I need to carry their energy in my womb, mind, or spirit."*

- *"I return their energy to them. I call my energy back to me."*

Step 3: Speak the Release

Hold the paper in your hand or place it between your palms at your heart center. Say aloud:

> *"I honor the lessons, and I release the energetic cords. I do not belong to them. They do not belong to me. I reclaim all parts of myself now. From this moment forward, I walk in freedom, sovereignty, and sacred self-possession."*

(Optional: burn the paper safely, or submerge it in the bowl of water to symbolize release.)

Step 4: Cleanse + Close the Portal

- Dip your fingers in the water or salt and gently anoint your womb space, heart, and forehead.
- Breathe deeply and visualize golden light filling your body.
- Imagine every space that once held their energy is now being sealed with light, love, and divine protection.

Say:

> *"I am whole. I am free. I am sovereign.*
> *I now close this portal with gratitude and power."*

Blow out the candles. Sit in silence. Let yourself feel what rises. Cry. Breathe. Smile. You're home.

Types of Energy Exchange During Sex

What You Give. What You Receive. What You Absorb.

Every sexual experience is an **energetic conversation**.
Your body may speak in breath and touch— but your **energy field** speaks in vibration, emotion, and intention.

Whether it's a one-night encounter or a long-term partnership, sex is not just a physical event. It is a merging of two (or more) energetic systems—each carrying their own histories, trauma, desire, and consciousness. When we're not aware of what we're exchanging, we absorb what was never meant to be ours.

1. Emotional Energy

What a partner feels during sex—love, resentment, neediness, control, grief—**doesn't stay in their body.**
It moves into yours. Especially during orgasm, when your energetic defenses are lowered, emotions can transfer silently into your system. If your partner is emotionally repressed, in pain, or projecting fantasies onto you...
you may leave the encounter feeling confused, low, or "off." Even if the sex was consensual. Even if you thought you were in control.

2. Trauma & Wounded Energy

Unprocessed trauma can leak through touch. If your partner has experienced abuse or holds internalized rage, guilt, or shame—it can become a **karmic residue** absorbed during sex. Their nervous system patterns may begin to affect your own. This is why some people develop anxiety, nightmares, sadness, or disconnection **after** sleeping with someone new. You weren't just with their body. You were with their *story*—their wounds, their ghosts, their unmet needs.

3. Spiritual Attachments + Entities

If someone is carrying spiritual baggage—addiction energy, jealousy, toxic intentions, or energetic parasites—they may unknowingly pass it on during sex. Especially if they've had multiple unprotected or unconscious sexual encounters, these entities can **attach** to your aura or womb space like spiritual hitchhikers.

Symptoms may include:

- Feeling spiritually heavy or "cloudy"
- Losing motivation or creative drive
- Strange dreams or increased spiritual sensitivity
- Obsessive thoughts about the person, even if you want distance

4. Creative & Life Force Energy (Sacral Exchange)

Sex isn't just about releasing tension—it's about **creating** energy. Your sacral chakra is your center of creativity, pleasure, and **life force**. When activated, it opens the portal for manifestation, inspiration, and spiritual power.

If sex is aligned and sacred, both people leave **energized, radiant, and creatively awakened**. If sex is misaligned, unreciprocated, or manipulative, one person may leave **depleted**, while the other leaves **fed**. This is often how energetic vampirism happens through intimacy.

> *Sex is not casual when you are conscious.*
> *And energy is not exchanged equally when intentions are unspoken.*

When you begin to **listen energetically**, you'll know when a person's touch brings peace— and when it brings poison.

Spiritual Hygiene After Sex

Protecting Your Aura. Purifying Your Energy. Returning to Yourself.

Just as we cleanse our bodies after sex, we must also cleanse our **energy.** Because sex, even when consensual and beautiful, opens portals—between bodies, between hearts, and between worlds. And what flows in during that time isn't always love.

You may leave the encounter with someone else's sadness, tension, confusion, or projections clinging to your field like smoke. That's not because you're weak. It's because you're **receptive**—and your body deserves restoration. Spiritual hygiene is how we **close the energetic portal** after sex. It's how we call our energy back. It's how we release what's not ours and reaffirm our sacred sovereignty.

Why Spiritual Hygiene Matters:

- It helps you ground and return to your own frequency
- It prevents energetic entanglement, obsession, or depletion
- It clears emotional or psychic residue left behind in your aura
- It allows you to integrate the experience (instead of absorb it blindly)
- It supports your nervous system in finding balance and peace

Simple Practices for Spiritual Cleansing After Sex

1. Salt Bath + Intention Ritual

Draw a warm bath with sea salt, Epsom salt, or pink Himalayan salt. Add herbs like lavender, rosemary, or rose petals.
Speak aloud:

> *"I release any energy that does not belong to me.*
> *I return to myself fully, cleansed and clear."*

Soak and imagine your aura being rinsed in golden light.

2. Smoke Cleanse or Aura Sweep

Use sage, palo santo, mugwort, or cedar.
Sweep the smoke around your body, especially your womb, heart, and crown. Visualize cutting invisible cords and sealing your field with light.

3. Sound + Frequency Clearing

Play healing frequencies (396 Hz, 528 Hz, or 963 Hz), or use singing bowls, chimes, or your voice. Hum or chant to break up lingering energy and harmonize your vibration.

4. Breathwork + Womb Massage

Lay down and place your hands over your womb or solar plexus. Take slow, deep breaths into your belly.
Say:

> *"This is my center. I reclaim my rhythm. I am safe inside myself."*

Use gentle massage to restore flow and reconnection.

5. Cord-Cutting Visualization

Envision any cords of attachment or leftover energy dissolving in light.
Say:

> *"I thank you. I release you. I reclaim me."*

End with a hand on your heart and womb.

Bonus Practice: Yoni Steaming or Womb Care

Use a gentle yoni steam with herbs like calendula, basil, mugwort, or rose. This supports not only physical cleansing but spiritual recalibration of your womb space—your seat of intuition and power.

Cleansing after sex isn't about guilt. It's about **integration.** It's a sacred return. To yourself. To your truth. To the wisdom of your energy field—whispering, *"Thank you for remembering me."*

Post-Encounter Care Is Intimacy, Too

The Sacred Afterglow: How We Love Each Other When the Fire Softens

The moments *after* sex are just as sacred as the ones before and during. Because what happens in the **afterglow** reveals the heart behind the heat.

Too often, couples collapse or turn away the moment climax is reached—forgetting that true intimacy lives in the space *after* release.

Post-encounter care is not an obligation.
It's a sacred offering.
It says:

> *"I'm still here."*
> *"You are not just a body—I see your soul."*
> *"You are safe in my presence, even when the fire dims."*

What Does Aftercare Look Like?

Aftercare doesn't have to be grand or performative.
It just has to be **intentional.** Here are ways partners can tend to each other's bodies, hearts, and nervous systems after

intimacy:

- **Ice water or fruit** to rehydrate after intense sessions
- **Showering or bathing together** to cleanse and reconnect
- **Offering warm washcloths or sanitary wipes** to gently clean sweat, fluids, or lingering touch
- **Holding, cuddling, or resting skin-to-skin** to support nervous system regulation
- **Whispering words of affection or gratitude**, like:

  "Thank you for trusting me."
  "You're safe. I'm here."
  "That felt beautiful. Are you okay?"
- **Breathing together** in silence to anchor the body and spirit back into presence

Why Aftercare Matters (Scientifically & Spiritually):

- It supports **oxytocin release**, strengthening emotional bonds
- It regulates the **parasympathetic nervous system** (calming after arousal)
- It affirms **safety**, especially for trauma survivors or highly sensitive people
- It seals the **energetic portal** that was opened during sexual connection
- It reinforces the message:

"I don't just want your body—I want your well-being."

When sex ends in silence or distance, it can feel jarring—like falling from a high without a net.

But when sex ends in **care**, in **tending**, in **touch that says "I still see you"**— it becomes a healing. A memory. A prayer. Because passion will fade. But *presence*? That's what we never forget.

Part II

Spirit & Sex – The Energetic Blueprint of Intimacy

Where Flesh Meets Frequency, and Pleasure Becomes Prayer

Sex is not just physical.
It is spiritual. Energetic. Alchemical.

It is a portal through which we remember:

> *We are not just bodies performing acts of passion. We are souls exchanging energy. We are creators in motion. We are divine in desire.*

In this section, we leave behind the scripts we've inherited from porn, patriarchy, and suppression—
And step into **the sacred**, where intimacy is a dance between energies, and the body becomes a vessel for manifestation, healing, and connection.

Here, you'll explore:

- The **womb as portal** and a map of your creative, intuitive power
- The **ritual of sacred sex**—from tantric breathing to energetic union
- The magic of **orgasm as manifestation**, intention, and liberation
- The unspoken **contracts and cords** we form through sexual exchange

- The need for **spiritual hygiene, boundaries, and energetic discernment**
- And how sex, when honored with intention, becomes a source of abundance, clarity, and cosmic power

This is not just about having better sex. This is about having **conscious sex**. Embodied sex. Aligned sex.

> *The kind that awakens—not just your senses—but your spirit.*

Chapter 5

The Womb as a Portal and Universe

Where Creation Begins—Literally and Spiritually

Before there was form, there was **void**. Before there was sound, there was **vibration**. And from that space of darkness, fluidity, and infinite potential—**creation** emerged.

That is the womb. Not just the physical organ—but the **energetic womb** that lives within every woman and feminine being. It is where life begins. Where dreams gestate.
Where energy is stored, processed, and released. It is the seat of our intuition, power, pleasure, and pain. And it is one of the most **spiritually charged** centers in the human body.

When we treat the womb only as a reproductive tool or a source of shame, we sever ourselves from our most sacred source of power. But when we return to it as a **portal**, a **temple**, and a **universe of its own**—we begin to reclaim ourselves on a soul-deep level.

The Physical Womb and the Energetic Field

The uterus is shaped like an inverted triangle—a powerful sacred geometry. It holds the blueprint for gestation, transformation, and birth.
But beyond the physical organ, the **energetic womb** holds:

- Old lovers and past traumas
- Unspoken grief, sexual shame, and ancestral wounds
- Creative ideas not yet birthed

- Psychic and emotional residue
- Spiritual codes waiting to be awakened

This is why some women feel "heavy" in their lower belly or disconnected from their sensuality—it's not just physical…it's **energetic overload**.

Womb Imprints: What We Carry and How to Cleanse It

Clearing Residue, Releasing Stories, Reclaiming Power

Your womb is more than biology. It is a **memory keeper**.
A spiritual library of every intimate encounter, every betrayal, every heartbreak, every birth, every loss, every moment you said "yes" when you meant "no."

Even if you've never had a child—your womb holds echoes of what was planted. Not just in this life…but through **ancestral lineage, cultural conditioning, and soul memory.**

What Lives in the Womb Unhealed?

- **Energetic residue** from past lovers or one-night stands
- **Shame** around pleasure, menstruation, or body image
- **Abuse or trauma** (sexual, emotional, spiritual)
- **Creative blockages** and unfulfilled desires
- **Abortions, miscarriages, or childbirth trauma**
- **Internalized messages** that sex is dirty, sinful, or only for others' pleasure
- **Ancestral pain**—grief passed down through generations of silence, suppression, or exploitation of

the feminine

This is why some women experience:

- Chronic pelvic tension or pain
- Disconnect from sexuality or sensuality
- Difficulty creating or expressing themselves
- A sense of numbness, fatigue, or emotional weight in the lower belly

These are not random symptoms. They are **imprints**. They are invitations.

Cleansing the Womb Space: A Sacred Return

You cannot reclaim your power without first creating space for it. This is not about purity—it's about **presence.**

To cleanse the womb is to say:

> *"I deserve to feel free in my body."*
> *"I am safe enough now to let go."*
> *"I am ready to hear the wisdom underneath the pain."*

Practices for Energetic Womb Cleansing:

Womb Journaling + Somatic Release

Write letters to past lovers, traumas, or versions of yourself you're ready to release. Then burn them safely—symbolizing a spiritual uncoupling.

Yoni Steaming

Use herbs like mugwort, lavender, rose, or calendula.
As steam rises, visualize it pulling out old energies, grief, or residue that no longer belongs.

Womb Breathing + Visualization

Lie down, place both hands on your lower belly.
Breathe in light, exhale anything heavy or lingering.
See your womb as a glowing orb—clearing, softening, awakening.

Chakra Alignment + Sound Healing

Use crystals (like carnelian or rose quartz), sound bowls, or affirmations to balance the sacral chakra and restore vibrational harmony.

Your womb was never meant to be a graveyard. It is a **garden.** And when you clear what was buried, you make room for what is meant to bloom.

You do not need to carry everyone you've loved, lost, or tried to save. You are allowed to release. You are allowed to rest. You are allowed to be **reborn**.

The Womb as Portal: Creation, Sex, and Divine Feminine Power

Awakening the Sacred Intelligence Within

The womb is not just where babies are born. It's where **worlds are built.** It is where ideas gestate, intuition is birthed, and soul codes awaken. It is where life force meets desire... and matter responds. It is the original **portal**—a gateway between realms, between ancestors and descendants, between spirit and flesh.

When you understand this truth, you stop treating your womb as something passive or shameful—and start treating it as a **living altar** of power.

Your Womb Is...

- A **channel** for divine downloads and creative visions

- A **record keeper** of sexual, emotional, and energetic experiences
- A **resonance chamber**—amplifying the energy you put into it
- A **cosmic gateway** between the spiritual and physical planes
- A **ritual vessel** for pleasure, presence, healing, and manifestation

This is why conscious sex is sacred...and unconscious sex can be spiritually devastating. Because what enters your womb doesn't just enter your body—it enters your *field.*

Pleasure as Portal

When a woman experiences pleasure—especially deep, intentional pleasure—her entire energy field **opens.**
Blood flows. Hormones shift. Emotions soften. Brainwaves drop into trance-like states.

And when orgasm happens, something magical unfolds:

- The **crown chakra** opens to receive divine light
- The **heart chakra** expands in unconditional love
- The **sacral chakra** activates creative potential
- The **womb space** begins pulsing with life force and spiritual frequency

This is not just climax. This is **activation.**

Why the Womb Is Key to Manifestation

The law of attraction is powerful. But **the womb is the original law of creation.** It doesn't just want thoughts—it wants embodiment.

You cannot just think your way into desire. You must *feel* it. You must *plant it.* You must let it take root in your body, and gestate in the mystery of your womb. Whether you are manifesting healing, abundance, love, or legacy—the womb is your portal. You are not disconnected. You've just been disempowered. But your womb has been waiting for you to remember...You are not just a woman.
You are a **universe.** And within you lives the **blueprint of creation itself.**

Womb Activation Affirmation

Speak this aloud. Let every word enter your body as truth.

> *I return to my womb with reverence.*
> *I honor it as my center of power, wisdom, and divine creation.*
>
> *This is no longer a place of pain or shame.*
> *This is a place of healing, of memory, of magic.*
>
> *I release what does not belong to me.*
> *I call back every part of my energy, my voice, my essence.*
>
> *I plant joy here.*
> *I plant vision here.*
> *I plant pleasure here.*
>
> *My womb is a portal of manifestation.*
> *My womb is a vessel of light.*
> *My womb is sacred—and so am I.*
>
> *I am whole.*
> *I am powerful.*

I am ready.

Chapter 6

Sacred Sex, Tantric Traditions, and the Power of Presence

Ancient Wisdom for Modern Pleasure

In a world addicted to speed, conquest, and performance—**Sacred Sex** is rebellion. It is remembrance. It is the slow, intentional art of *becoming God through touch.*

This isn't about quick orgasms or perfect techniques. This is about **presence**. About awakening the divine within yourself and your partner through breath, eye contact, devotion, and **energetic attunement**.

Tantra, one of the oldest spiritual systems in the world, teaches us:

> *Sex is not separate from the sacred. It is the sacred. When honored with awareness, sex becomes meditation, medicine, and manifestation.*

This chapter is a reclamation. Of ancient wisdom. Of sensual intelligence. Of the truth that sex—done consciously—can open portals of love, healing, ecstasy, and even divine union.

What Is Sacred Sex? (And What It Is Not)

Reclaiming the Divine in Desire

Sacred sex is not about performance. It's not about proving your worth, mastering a technique, or chasing climax.

Sacred sex is **presence with purpose.**

It's devotion with your body. It's choosing to see your partner —not just as a lover, but as a reflection of the divine.

It is not a separate act from your spirituality.
It *is* your spirituality—in motion.

What Sacred Sex *Is:*

- A practice of **conscious connection**, where both people feel safe, seen, and sovereign
- A space where breath, intention, and emotional presence matter as much as touch
- An invitation to **merge energetically**, not just physically
- A slow, spacious unfolding—not a rushed race to orgasm
- A **healing portal** for the nervous system, inner child, and heart
- A moment where your soul and your body are saying **yes** together

What Sacred Sex *Is Not:*

- It is not performative—you don't have to moan a certain way or look a certain way
- It is **not self-sacrificial**—your needs, boundaries, and pace matter
- It is **not rooted in shame, guilt, or obligation**

- It is **not about "fixing" or "keeping" someone** through sex

- It is **not bypassing** pain with pleasure or using intimacy to distract from your truth

Sacred Sex Is a Frequency

It doesn't require candlelight or crystals (though you're welcome to include them). It requires **awareness**.

It requires **truth**. It requires the courage to **feel, pause, and be fully present**—with yourself, and with your partner. It is a ceremony. A communion. A conversation between spirit and skin.

Before we explore Tantra as one of the ancient blueprints of sacred sexuality, remember this:

> *Sacred sex is not about how much you do…*
> *It's about how much of yourself you're willing to bring to the moment.*
>
> Presence is the portal. Pleasure is the offering.
> And love—true, conscious love—is the altar.

Tantra: Origins, Misconceptions & Spiritual Foundations

Returning to the Source of Sacred Union

The word **Tantra** comes from the Sanskrit root words *"tan"* (to expand) and *"tra"* (to liberate).
It literally means:

> **"To weave" or "to expand beyond limitations."**

Tantra is not just about sex—it is a **spiritual path**.
A practice of total presence. Of merging masculine and feminine energy, within and without. Of seeing *everything*—

the body, the breath, the beloved—as a portal to the Divine.

Originating over 5,000 years ago in India, Tantra was a **mystical science** practiced by sages, yogis, and healers. It was rooted in:

- **Breathwork** (Pranayama)
- **Meditation**
- **Sacred touch**
- **Mantra and sound**
- **Chakra alignment**
- **Energetic balancing of Shiva (masculine consciousness) and Shakti (feminine life force)**

What Tantra *Is Not*:

- It is not about *orgies* or wild sexual acts
- It is not just for couples or heterosexual unions
- It is not about bypassing emotion or trauma for pleasure
- It is not a sexual technique to "last longer" or "please better"
- It is not performance-based or rooted in ego-driven conquest

Western culture has distorted Tantra into a buzzword for erotic novelty. But true Tantra is sacred. It is subtle. It is the path of awakening **divinity through the body**, not escaping it.

At Its Core, Tantra Teaches:

- Your **body is not separate** from your spiritual path—it is a temple
- **Pleasure is a gateway**, not a distraction
- **Presence is the highest form of foreplay**
- Every breath, every gaze, every sacred touch is a step toward liberation
- The union of two people can mirror the **cosmic dance** of masculine and feminine energy within creation itself

Tantra is the invitation to slow down. To listen. To open.
To experience yourself—and your partner—not as objects... but as **gods in human form**.

Core Practices of Tantric Intimacy

Breath, Eye Gazing, and the Power of Energetic Presence

Tantra is not about adding more—It's about *removing* what disconnects us. When you strip away the noise, the performance, and the pressure...what's left is pure presence. And it is from that presence that **Tantric intimacy** emerges. You don't need years of training to begin. You only need willingness. To slow down. To feel more. To *listen* with your whole being.

1. Breathwork – The Foundation of Tantric Connection

Your breath is your life force. In Tantra, it's the bridge between your physical body and your energetic body. Breathing together creates **synchronization**—a sacred rhythm where your hearts, nervous systems, and energy fields begin to **entrain** and harmonize.

Practice:

- Sit or lie facing your partner.
- Inhale slowly together for 4 counts, exhale for 4 counts.
- Breathe in through the nose, out through the mouth.
- Imagine a loop of breath flowing between you—cleansing, connecting, energizing.

When breath slows, the soul speaks louder.

2. Eye Gazing – Seeing the Divine in Each Other

In Tantric practice, the eyes are more than sensory organs—they are **gateways to the soul.** Eye gazing dissolves ego, reawakens emotional connection, and activates the heart chakra. It can be deeply vulnerable. And deeply healing.

Practice:

- Sit comfortably, facing each other.
- Maintain soft eye contact without speaking for 2–5 minutes.
- Breathe slowly. Stay open. Let whatever emotions arise come and go.

You may begin to feel waves of love, laughter, tears, or even resistance. That's okay. Keep breathing. Keep seeing. Let the walls fall.

To truly be seen—and to stay—is one of the most sacred erotic acts.

3. Energy Work – Moving Pleasure Beyond the Physical

In Tantra, touch is not mechanical—it is **intentional, reverent,**

and alive. You are not just touching a body. You are touching a being of light.

Energy work during intimacy can include:

- **Hovering hands** just above the skin to sense energy fields
- **Light fingertip tracing** along meridians (neck, spine, inner thighs)
- **Holding pressure points** while breathing together
- **Chakra-to-chakra connection**, such as heart to heart or sacral to sacral
- **Sound and vibration** to release blocked energy (moaning, humming, chanting)

The goal is not climax. The goal is to **raise energy**, circulate it, and allow the whole body to become a vessel of awakening.

> *Your touch can heal. Your presence can transform. Your intention is the spell.*

Tantric intimacy is less about technique...and more about **tuning in**.

When you honor breath, presence, and energy, you create a space where souls speak, bodies open, and love becomes **transcendent**.

Tantric Ritual Flow for Beginners

A Sacred Sequence to Awaken Connection, Energy & Divine Presence

> *This ritual isn't about "having sex."*
> *It's about deepening intimacy—through breath, touch, and presence. It's a ceremony where both bodies become altars, and the union becomes divine.*

You don't need to be experienced in Tantra to begin.
You only need:

- A quiet space
- An open heart
- Time without rushing
- And the willingness to be fully present with yourself and your partner

Set the Container (5–10 mins)

1. Prepare the space

- Light candles or incense
- Clear clutter or distractions
- Set an intention together (whisper it or write it down)

2. Create a shared grounding moment

- Sit or stand facing each other
- Place your left hand over your heart, your right hand on your partner's heart

- Breathe together for 2–3 minutes
- Silently affirm: *"I honor the divine in you as I honor the divine in me."*

Synchronize Your Breath (5 mins)

3. Practice circular or mirrored breathing

- Inhale together for 4 counts, exhale for 4
- Option: One partner inhales while the other exhales—creating a loop of energy
- Maintain gentle eye contact
- Feel your nervous systems begin to entrain and relax

Eye Gazing + Energetic Alignment (3–5 mins)

4. Soft gaze into each other's left eye (the intuitive/spiritual eye)

- Observe, don't analyze
- Let any laughter, discomfort, or emotion flow—this is part of intimacy
- Optional: Place palms together in front of your hearts (Anjali Mudra)

This is the moment the veil drops. Let your soul speak without words.

Sacred Touch + Sensual Mapping (10–20 mins)

5. Explore touch without expectation

- Use fingertips, breath, and energy to trace the body—not just genitals
- Focus on erogenous zones: neck, shoulders, back, inner thighs, lower belly
- Ask: "Do you want more pressure? Less? Faster? Slower?"

6. Optional: Chakra-to-Chakra Alignment

- Lie face-to-face or sit in Yab Yum (one partner on the other's lap, heart to heart)
- Breathe into the chakras together—starting at the root, up to the crown
- Visualize golden or rose light connecting each center

Closing the Ritual (5–10 mins)

7. Seal the connection with gratitude

- Hold each other or rest with hands over each other's hearts or wombs
- Speak a word of gratitude or reverence

Tantric rituals are **living prayers**. Let them evolve with you. Make them your own. Because every time you slow down, breathe deeper, and connect from your heart—you are practicing Tantra.

> "Thank you for showing up."
> "You are safe with me."
> "I see the divine in you."

8. Rest in stillness or take a cleansing bath together

- Let energy settle

- No rushing. No pressure. No expectation.

This isn't the end. This is integration.

Reflection + Integration Journal Page

After the Ritual: Listening to the Wisdom of Your Body, Heart & Energy

You've just shared something powerful. Now it's time to reflect —not with judgment, but with love. This page is a space to witness yourself and your partner with honesty, compassion, and curiosity.

Take a few deep breaths. Let the experience settle. Then write freely.

What emotions came up for me during the ritual?

How did I feel in my body throughout the experience?

☐ Grounded
☐ Awakened
☐ Numb
☐ Anxious
☐ Safe
☐ Open
☐ Disconnected
☐ Held
☐ Turned on
☐ Emotional

Other sensations or body messages:

What part of the ritual felt the most intimate or powerful for me?

What did I notice about my partner during the experience?

What did this ritual teach me about myself or my relationship to intimacy?

Is there anything I want to explore more deeply next time?

Final Reflection:

In one word, how do I feel now?

What is one thing I want to remember from this ritual experience?

Closing:

Place your hand on your heart and say aloud:

"I honor the version of me that showed up today. I trust the wisdom of my body. I am open to deeper connection, guided by truth, love, and presence."

The Power of Slowness, Surrender & Non-Goal-Oriented Pleasure

Where the Body Opens, the Spirit Awakens

In a world obsessed with quick fixes, fast orgasms, and pornographic perfection—choosing **slowness** is revolutionary.

To slow down is to **listen.** To give the body time to speak.
To allow pleasure to rise in waves, not in deadlines. To discover the infinite sensations available *between* the beginning and the end.

Slowness Is a Portal

When you slow down, you shift out of:

- Performance
- Pressure
- Numbness
- Anxiety
- Outcome obsession

And you return to:

- Breath
- Presence
- Trust
- Embodiment

- Reverence

Slowness allows the body to soften. It invites the nervous system to settle. It gives space for emotional intimacy, safety, and sensual wisdom to emerge.

What if your goal wasn't to "finish"—but to feel?

Surrender Is Not Weakness—It's Receptivity

In sacred sex, surrender doesn't mean giving away your power. It means choosing to be **open**—to yourself, to your sensations, to your truth.

Surrender allows you to:

- Receive without guilt
- Let go of control and enter the unknown
- Trust your partner, your body, and the present moment
- Experience deeper levels of orgasm, energetic merging, and emotional release

To surrender is to say, "I am safe enough to be seen. I am free enough to be felt."

What Is Non-Goal-Oriented Pleasure?

Non-goal-oriented pleasure is pleasure without a finish line. It's not about reaching orgasm—though that may happen. It's about **exploration, intimacy, and communion.**

This kind of pleasure:

- Honors **every sensation** as sacred
- Allows **playfulness, curiosity, and freedom** to emerge

- Invites **multiple waves of pleasure**, not just a singular climax

- Expands your capacity to **stay present** instead of chasing results

Pleasure is not a destination. It's a vibration.
And when you let go of the "goal," you discover the goddess within the moment.

The most powerful sex isn't always the loudest or wildest. It's often the quietest. The slowest. The kind that makes you weep, not because it hurt—but because you felt so seen.

Sacred sex asks you to **come undone** without fear. To let the clock melt. To let love lead. Because the body knows. And when you trust its rhythm, you remember that **pleasure is your birthright—not your performance.**

Chapter 7

Sex Magic and Manifestation

Harnessing Orgasm as a Force for Abundance

There is a reason ancient priestesses used sex in ritual. There is a reason spiritual initiates practiced celibacy or ecstatic pleasure in cycles. There is a reason you feel most alive, open, and magnetic after deep, conscious intimacy.

Because sex is **creation energy**. And when it is paired with clear intention, it becomes **magic**.

Sex magic is not about control. It's not about manipulation or seduction. It's about aligning your energy, body, breath, and desire to **co-create with the Universe.**

Orgasm is the highest natural energetic spike the human body can access— a moment of surrender, expansion, and transcendence. At the peak of climax, your brain releases waves of dopamine, oxytocin, endorphins, and nitric oxide— but spiritually, your field becomes **wide open** to receive and transmit intention.

> *What you feel during orgasm, you program into your field. What you focus on during orgasm, you invite into form.*

The Neuroscience of Manifestation During Orgasm

At climax, the brain undergoes a brief but intense alteration in consciousness:

- **Dopamine** floods the brain, creating feelings of reward and heightened focus.

- **Oxytocin** expands connection and opens the heart field —one of the most powerful magnetic frequencies we emit.

- **Theta brainwaves** may emerge, which are associated with deep meditation and hypnotic receptivity.

- The **default mode network** of the brain (linked to ego and overthinking) shuts down—creating a sense of unity, transcendence, or "oneness."

- The **limbic system**, which processes memory and emotion, is activated—meaning whatever you're feeling or thinking at climax gets stored as a powerful energetic and emotional memory.

In that moment, you are not just releasing—You are ***imprinting.*** *You are casting a spell through biology and spirit.*

The Quantum Physics Behind Orgasmic Intention

From a metaphysical lens, orgasm generates a spike in **vibrational frequency**.

In that heightened state:

- Your **energetic field expands**

- Your **intention becomes magnetic**

- Your body acts as a **living antenna**, broadcasting that frequency into the quantum field

- The Universe, which responds not just to words but to **vibrational alignment**, begins to reorganize reality to match the energetic "request" you've made

This is why focusing on fear, resentment, or self-judgment during sex can actually create more chaos in your field—while pleasure infused with clarity, love, and intention becomes a powerful force for manifestation.

Orgasm Is a Portal. But Intention Is the Key.

Sex magic is not about forcing outcomes. It's about **alignment**. It's about knowing your body is divine technology—designed to co-create with Source.

When you combine orgasmic energy with spiritual awareness and emotional clarity, you become an unstoppable manifestor. You don't have to be perfect. You just have to be **present.** And willing to believe that your pleasure is a valid, sacred, and effective path to abundance.

When Pleasure Becomes a Distraction:
The Manipulation of Orgasm

Orgasm is powerful. It alters your brain chemistry. It opens your body. It quiets your mind. But that power can also be **misused.**

In many relationships, sex becomes a substitute for resolution — a momentary balm for deep emotional wounds that remain unspoken.

The Neurochemistry of Suppression

During orgasm, the brain floods with:

- **Dopamine** (reward, motivation)
- **Oxytocin** (bonding and trust)
- **Endorphins** (pain relief, euphoria)

These chemicals create a **temporary state of peace, connection, and contentment**. So it can feel like:

- "We're okay again."
- "I forgive you."
- "Let's just move forward."

But in truth...you may not have resolved anything.

You simply overrode your pain with pleasure.
You buried the issue beneath a neurochemical high.

Make-Up Sex, Misdirection & Emotional Avoidance

It's not uncommon for couples who argue—sometimes even aggressively—to end up in bed, claiming "passion" as a resolution.

But was it connection? Or **convenient suppression**?

Over time, when issues are repeatedly swept under the rug through sex, a few things happen:

- Communication weakens
- Emotional intimacy erodes
- Resentment builds beneath the surface
- The body begins to protest—through numbness, shutdown, or avoidance
- Eventually, even sex can't cover the wound

You can't orgasm your way through emotional avoidance forever. What's not expressed will eventually erupt—or disappear completely.

Sex as a Tool for Control or False Bonding

Some people, consciously or unconsciously, use sex to manipulate:

- To regain power after conflict
- To avoid vulnerability or apology
- To bond someone to them through orgasmic intensity
- To create soul ties they have no intention of honoring emotionally or spiritually

This is why we hear things like:

> *"Why is the sex so good with toxic people?"*
> *"Why do I feel more addicted to them after we sleep together?"*

It's not because it's **healthy**. It's because it's **high.** Because their only channel of emotional expression is through physical release. They don't *connect* through conversation, truth, or soul—but they've mastered how to **stimulate your nervous system.**

Sex without emotional safety can feel electric—but it is a **false fire.** You might burn brightly in the moment, but you will burn out in the long run.

Sacred sex isn't about avoiding the hard truths— it's about creating a safe enough space to bring them forward...and still choose connection.

Because the best sex isn't found in chaos. It's found in *clarity.* And clarity is only possible when you're no longer using sex to forget what your soul needs you to remember.

How Men & Women Use Sex Magic Differently

Understanding Polarity, Power, and the Energetics of Creation

While sex magic can be practiced by **anyone**, regardless of gender identity, it's important to understand that **masculine and feminine bodies** (as well as energies) move differently through orgasm, intention, and energy release.

This is not a hierarchy—this is a **harmony**. Each carries distinct strengths, sacred roles, and energetic rhythms in the dance of divine creation.

The Masculine Principle (Seed, Penetration, Intention)

In the sex magic dynamic, the **masculine** energy represents:

- **Direction and focus**
- **Penetrative intention** (both physically and spiritually)
- **Holding the container of consciousness**
- **Providing safety for surrender to occur**

For men or those embodying masculine energy:

- Orgasm often results in a release or **discharge of energy**
- Sex magic is most potent when they channel their orgasmic force with *clarity and restraint*
- Retaining energy (through breath, control, or even non-ejaculatory orgasm) allows them to **build magnetic force** rather than deplete it
- The power lies in **where the mind is during climax**—whatever is focused on will be *projected outward into the*

field

Masculine sex magic is about focused intention, protection, and sacred offering.

The Feminine Principle (Womb, Reception, Amplification)

The **feminine** energy in sex magic represents:

- **Receptivity and expansion**
- **Womb-space as the portal of manifestation**
- **Emotional depth and alchemical transmutation**
- **Energetic amplification and magnetism**

For women or those embodying feminine energy:

- Orgasm opens the **energetic and auric field** wide—making the moment ideal for calling in desire
- The womb acts as a **magnetic cauldron**, where intention gestates and grows
- Feminine energy draws inward to **receive, nourish, and magnetize** what is desired
- Her power is in *feeling deeply*, not forcing—because what she feels at orgasm becomes a **vibrational command**

Feminine sex magic is about receptivity, embodiment, and sacred magnetism.

When Both Energies Unite

When conscious masculine and feminine energies meet in sacred sexual union—whether in one body or between two—it becomes a cosmic ritual:

- The **masculine provides the intention and direction**
- The **feminine receives, amplifies, and births that intention into form**
- Both surrender to the divine current moving through them
- Their unified energy creates a **vibrational resonance** that can ripple into physical manifestation

This is the sacred blueprint of co-creation. This is divine sex. This is the tantric union of **Shiva and Shakti**, heaven and earth, seed and soil.

You don't need a partner to practice sex magic. But when polarity is honored—whether within yourself or in union with another—your power **multiplies**.

The masculine penetrates with clarity. The feminine receives with trust. Together, they **create worlds.**

Sex Magic Ritual Guide

Solo & Partnered Practices to Manifest with Pleasure

You are the altar.
Your intention is the spell.
Your orgasm is the ignition.

Sex magic is not about performance—It's about presence.
It's not about how long or how loud—It's about how **aligned** you are in thought, energy, and feeling.

These rituals are designed to help you harness orgasm as a sacred, creative force. Whether solo or with a partner, your pleasure becomes a powerful act of **intentional manifestation**.

Before You Begin: Set Your Sacred Space

- Cleanse the room (smoke, sound, water, prayer)
- Dim the lights, light candles or incense
- Choose a sensual playlist if desired
- Optional: place crystals (rose quartz, carnelian, citrine) or a journal nearby
- Prepare your **intention**—a clear, heart-aligned desire (health, love, wealth, clarity, etc.)

Sex magic begins long before touch. It begins in ***devotion.***

Solo Ritual: The Womb + Will Manifestation Flow

1. **Ground + Center (5 min)**

- Sit or lie down with one hand on your heart, one on your womb or genitals
- Breathe slowly. Feel your body.
- Whisper: *"I call all of my energy back to me now."*

2. **Activate the Intention (3 min)**
    - Close your eyes and visualize what you desire *as if it's already yours*
    - Breathe it into your body—see it, feel it, smell it, embody it
3. **Pleasure Portal (10–30 min)**
    - Begin to touch yourself slowly and reverently
    - Focus on sensation—not rushing
    - Breathe deeply, moan, sound—allow your body to become the channel
4. **Climax + Release (Sacred Peak)**
    - As you feel orgasm build, return your focus to your intention
    - At climax, hold the vision.
    - Whisper it, cry it, shout it—*"It is done."*
5. **Integration (5–10 min)**
    - Lie in stillness. Let your body rest.

- Journal what came through.
- Thank your body for being the vessel.

Partnered Ritual: Unified Pleasure, Unified Power

1. **Create the Ceremony**
    - Face each other. Breathe together.
    - Share your intention aloud—or write it down and exchange.
    - Eye gaze. Touch gently. Create a sacred mood.
2. **Build Energy with Presence**
    - Engage in slow foreplay—without rushing to penetration or orgasm
    - Use sensual breathwork, sound, and synchronized movement
    - Whisper affirmations over each other like:

      "You are divine."
      "We create with love."
      "Together, we are magnetic."
3. **Conscious Climax**
    - As orgasm nears, focus your combined energy on a shared or individual intention
    - Visualize it together. Speak it aloud or silently

  - At climax, hold hands or place hands on each other's hearts or wombs
  - Let the intention pulse through both bodies

4. **Close the Portal**
  - Rest skin-to-skin.
  - Breathe and integrate together.
  - Thank each other for the sacred offering of presence and power.

Tips for Powerful Manifestation

- Avoid shallow or disconnected climax—it dilutes the energy
- Focus on how the **manifestation will feel**, not just how it will look
- Repeat your ritual with consistency, reverence, and joy
- Trust that **the Universe responds to your frequency, not your force**

You don't have to suffer for your desires. You can **pleasure** them into form. You can create from joy, from softness, from orgasm. Because your body isn't just divine—it's a **portal** of divine power.

Chapter 8

Blood Bonds, Rituals, and Sexual Contracts

What Happens Spiritually When We Merge

Sex is never just physical. It is emotional. It is energetic. And—whether you are conscious of it or not—it is **spiritual**. Every time we engage in intimacy, we create an **energetic exchange**. But in some instances, that exchange becomes a **contract**—a spiritual agreement, sealed not by paper, but by energy, bodily fluids, intention, and ancestral imprint.

In this chapter, we will explore:

- What blood bonds are and how they are formed
- Why sex during menstruation or shared blood rituals intensify energetic merging
- How spiritual contracts are created through sex, and what they mean
- How trauma, manipulation, or unspoken intentions can create **karmic entanglements**
- What it means to break or cleanse these energetic ties
- And why many people feel "haunted" by certain sexual experiences years after they've ended

You are not just sharing your body...

You are sharing your story, your spirit, your soul memory.

In some traditions, **the sexual act is seen as a ritual in and**

of itself. Every moan is a mantra. Every orgasm a rite. Every exchange—an imprint on your energetic field.

This chapter is not meant to instill fear. It is meant to offer **awareness**. Because once you understand the power of the portal—you begin to choose who you open it to... with discernment, with sovereignty, and with love.

Blood Bonds: Menstrual Sex, Blood Rituals & Ancestral Portals

The Sacred, the Sensual, and the Spiritually Binding

In nearly every ancient tradition, blood was considered holy. It was used to **seal covenants**, mark sacred rites, anoint altars, and connect with the spirit world. Why? Because **blood carries life force.** It is the physical expression of your soul's vitality—your ancestry, your essence, your spiritual DNA. When blood is involved in sex—whether intentionally or not—it becomes more than an act of pleasure. It becomes a **ritual**. It becomes a **binding.**

The Power of Menstrual Blood

Menstrual blood is not dirty.
It is **sacred**, potent, and energetically charged with:

- Emotional and hormonal frequencies
- Ancestral memory
- Womb intelligence
- Endings, release, and renewal cycles

When sex occurs during menstruation:

- The **womb is more open**, energetically and emotionally

- The **vaginal walls are thinner**, allowing for deeper absorption of energy

- The **emotional body is heightened**, which can increase psychic and energetic bonding

- If the partner is penetrating, their energy is **literally absorbed into the blood**—a deeply intimate exchange

Some practitioners honor sex during menstruation as a **ceremony of shedding and rebirth**. But when done unconsciously, it can also create a **blood bond** that is difficult to energetically sever.

A Scientific and Sacred Perspective

Menstrual blood has long been viewed through a lens of taboo, secrecy, and even shame. But in truth—**it is one of the most sacred and biologically rich substances a woman's body produces**. While many cultures celebrate menstrual blood as a potent source of power, healing, and spiritual insight, modern science also reveals its complexity, vulnerability, and the importance of honoring the body during this time.

What *Is* Menstrual Blood, Really?

Contrary to popular belief, menstrual blood is not simply "blood." It's a **biological cocktail** composed of:

- **Shedded endometrial lining**

- **Cervical mucus**

- **Blood plasma**

- **Unfertilized egg**

- **Immune and stem cells**

- **Inflammatory markers and enzymes**

Menstruation occurs when pregnancy does not take place. The hormone progesterone drops, signaling the uterus to shed its lining, which is then released through the vagina. This cyclical process is deeply tied to a woman's endocrine system, reproductive health, and energetic rhythms.

Clinical Source:
Saladin, K. S. (2012). *Anatomy & Physiology: The Unity of Form and Function*. McGraw-Hill Education.

Every Woman's Cycle is Unique

The *duration*, *volume*, and *consistency* of menstrual blood varies due to:

- Hormonal fluctuations
- Birth control
- Nutrition
- Stress and trauma
- Uterine health (e.g., fibroids, endometriosis, PCOS)

Heavy bleeding or **painful periods** can often signal underlying conditions, and intercourse during menstruation may exacerbate these if not approached with informed care.

When the Womb Is Wide Open—Understanding Vulnerability During Menstruation

From a **biological and spiritual** standpoint, a menstruating woman is *open*—energetically and physically.

Physiological Vulnerability

During menstruation:

- The **cervix may open slightly** to release the endometrial lining.
- The **vaginal pH** becomes more alkaline, increasing susceptibility to bacterial infections.
- **Immunity in the reproductive tract temporarily lowers**, making it easier for STIs or pathogens to enter the body.
- **Inflammatory cytokines** are elevated, contributing to cramps or discomfort.

Intercourse during menstruation—especially without proper hygiene, protection, and mutual consent—can increase risk of:

- **Pelvic Inflammatory Disease (PID)**
- **HIV and STI transmission**
- **Disrupted flora balance**
- **Secondary infections or inflammation**

Clinical Sources:

- Wira, C. R., & Fahey, J. V. (2008). The innate immune system: gatekeeper to the female reproductive tract. *Immunology*, 111(1), 13–15.
- Haddad, S. N., & Gilstrap, L. C. (2002). Infectious diseases during pregnancy. *Obstetrics and Gynecology Clinics of*

North America, 29(3), 541–550.

Misunderstood Risk: Transmission & Exchange

If either partner has a **sexually transmitted infection**, the exchange of blood increases the likelihood of passing on the infection—especially **HIV**, **HBV**, **HBC**, **HPV**, **chlamydia**, and **gonorrhea**. Blood is one of the most potent bodily fluids in disease transmission, and menstrual intercourse (without barriers like condoms) **amplifies the exposure risk**.

From a Spiritual Lens: Womb Blood is Sacred

While science validates the physiological realities, many indigenous and ancestral traditions speak of menstrual blood as:

- A **purifier**—cleansing the womb of what no longer serves.
- A **ritual offering**—used in earth-based spirituality for grounding, fertility rites, and healing.
- A **portal**—representing the life-death-rebirth cycle encoded within feminine wisdom.

In this phase, a woman's **intuition heightens**, her energy turns inward, and her boundaries may need strengthening. The womb is not just bleeding—it is **communing with the Divine**.

Consent, Communication, and Care

If a couple chooses to engage in intimacy during menstruation, it should be:

- Mutually consensual
- Clean and intentional

- Centered in reverence—not shame or secrecy

Practices like:

- Showering beforehand
- Using towels or waterproof barriers
- Practicing non-penetrative touch or oral boundaries
- Checking in emotionally and physically

...can all support safety and deepen connection.

Respecting the Red

Menstruation is **not dirty**, but it is **powerful**—
and with power comes responsibility.

Whether you choose to engage sexually during this time or not, it's essential to **honor the body's wisdom, vulnerability, and sovereignty**. The womb speaks loudly during this season. The question is: Are we listening?

Ancestral Portals: When Blood Awakens Lineage Memory

Your blood carries your lineage. When it is released during sex, it can awaken **ancestral codes, grief, or gifts**—both yours and your partner's.

Some symptoms of unhealed or unconscious blood bonding:

- Feeling inexplicably tethered to someone even after the relationship ends
- Recurring dreams or obsessive thoughts

- Loss of energy, motivation, or clarity post-encounter
- Intuitive "hauntings" or energetic weight in the womb
- Feeling like part of you was left with them—or they with you

You are not fragile for feeling deeply after sex involving blood. You are sensitive. You are spiritually intelligent.
And now, you are aware.

Blood is **not just fluid.**
It is **frequency.**
It is **portal.**
It is **power.**

And when you honor that—your body becomes not just sacred...but *sovereign.*

Why You Never Forget Your First Time: Blood, Trauma & Womb Memory

When the First Wound Becomes an Energetic Chain

We often hear, *"You never forget your first time."*
But for many people—especially women—that phrase doesn't come with nostalgia. It comes with discomfort, confusion, or pain.

What we don't often talk about is **why** that first sexual experience leaves such a powerful imprint—especially if there was **bleeding, pressure, pain, or trauma** involved.

It's not just about the act itself. It's about the **energetic coding** that happened in that moment.

The Womb as a Recorder of Firsts

Your womb is not just physical.
It is **an energetic archive**.

It remembers:

- How safe (or unsafe) you felt
- Whether the experience was loving or forced
- Whether your voice was honored—or silenced
- Whether your body opened from joy—or tensed from fear

And when the experience involves **blood**—from penetration, tearing, force, or menstruation—that imprint becomes even **deeper**. Why? Because **blood is a binder.**
It records the emotion of the moment. It seals it.
And if not released or healed, it **anchors that energy into the womb space**—where it can continue influencing our relationship with sex, love, and self-worth.

When Sex Becomes a Wound

For survivors of sexual trauma or abuse, especially where bleeding occurred, the body often registers that moment as a form of **energetic invasion**—not connection.

What may follow includes:

- Vivid memories or flashbacks
- Disassociation during intimacy
- Emotional shutdown
- Chronic pelvic pain, tightness, or numbness
- Difficulty forming or maintaining healthy sexual relationships

- Deep shame, guilt, or fear that arises during arousal

This isn't weakness. This is your body **remembering what your mind tried to forget**.

Why It Can Be So Hard to "Move On"

If the bleeding created a bond—however unwilling or unconscious—it can feel like part of you was **taken, held, or left behind.** Even if you never saw that person again. Even if you tell yourself it's in the past. Until those **energetic threads are cut**... until the womb is **cleansed and honored**...until your voice and body are **restored to you**—those imprints remain.

Not because you're broken. But because your body is wise. And it's still waiting for *resolution*. Your first time doesn't have to define you. Your trauma doesn't have to hold you. But your body needs a voice, a ritual, a release.

Because the womb doesn't forget.
But she can forgive.
She can cleanse.
She can be reborn.

And when she is, you become **free**.

Soul Ties & Energetic Imprints: How We Carry Our Lovers With Us

Unseen Threads, Unspoken Attachments, and the Echoes of Intimacy

Every sexual encounter leaves an **energetic fingerprint** on the body. Not just in the mind or in memory—but in the womb, the heart, and the auric field.

This is what we call a **soul tie**.

It's not just about love. You can create soul ties with people you barely know. With people you didn't want to connect to. Even

with people you tried to forget. Because sex is an **exchange of essence**—not just fluids or friction.
And once someone has entered your body, or your field, or your energy...a **thread** is formed.

There is a saying among spiritual healers: *"The womb remembers what the mind forgets."*

In every act of intimacy, a silent exchange takes place. Not just of pleasure, but of energy. Of emotion. Of essence. Of cellular codes. Sex is not casual. It is chemical.
It is karmic. And for women, especially, it is deeply **imprinting.**

Spiritual Insight: The Womb as a Portal and Storage Vessel

In metaphysical traditions around the world—from African ancestral healing to Taoist sexual alchemy—the womb is seen as:

- A **portal of creation and death**
- A **recorder** of experiences and energy
- A **receptacle** for both life force and pain

Women are energetically open and **receiving** during intercourse. The act of penetration is not just physical—it is spiritual and emotional entry. When ejaculation occurs inside the body, this sacred space becomes imprinted not just with seed—but with **memory**.

Scientific Findings: Can Male DNA Stay in a Woman's Body?

Modern science is beginning to uncover evidence that *the womb does not forget.*

One of the most referenced studies in this domain is from the **University of Seattle and the Fred Hutchinson Cancer Research Center** (Nelson et al., 2012), which found:

Male DNA can be detected in women's brains —even decades after sexual intercourse or pregnancy.

This phenomenon is part of a broader field called **microchimerism**, where foreign cells—often from a fetus, or in some theories, from semen—are absorbed and retained by the female body. These cells can lodge in various tissues, including the **brain, womb, liver, and bone marrow**.

Study Reference:
Nelson, J. L. et al. (2012). "Male Microchimerism in Women Without Sons." *PLoS ONE*, 7(9): e45592.
https://doi.org/10.1371/journal.pone.0045592

What This Might Mean for Women's Bodies

- These retained **male cells may influence gene expression** in the woman's body over time.
- They may also **interact with her immune system**, nervous system, and hormonal response.
- The presence of these foreign DNA fragments in brain tissue has raised questions about **long-term psychological or emotional effects**—a biological correlate to the spiritual concept of "soul ties."

While research is still early, it invites us to consider:
What are we absorbing through intimacy—beyond just pleasure?

The Psychological Imprint: Emotional Transfer and Personality Shifts

Women often report:

- Taking on their partner's **mood**, stress, or anxieties

- Adopting **language patterns**, habits, or even interests that weren't previously present
- Feeling **depleted** or "not themselves" after being with a toxic or emotionally volatile partner

This may be attributed to **neural co-regulation** and **limbic resonance**—where our nervous systems synchronize with those closest to us.

Clinical Insight:

- When sexual intimacy is combined with **repeated emotional engagement**, the brain forms **long-term neurological maps** and patterns of attachment—even if the relationship is unsafe.
- The **mirror neuron system** in the brain reflects the emotional states of intimate partners, often without conscious awareness (Keysers & Gazzola, 2009).

What Is a Soul Tie?

A soul tie is an energetic cord that forms between two people through:

- Sexual intimacy
- Deep emotional vulnerability
- Spiritual connection
- Or traumatic bonding

These cords can be:

- **Healthy** – rooted in love, respect, and mutual growth

- **Neutral** – momentary, light, and naturally dissolving over time

- **Toxic or traumatic** – rooted in pain, manipulation, abuse, or abandonment

Not all ties are meant to last. But when they're not acknowledged or consciously released—they linger.

Signs You're Carrying a Soul Tie or Imprint:

- You can't stop thinking about someone—even if the relationship ended long ago

- You feel drained, confused, or emotionally pulled when you think of them

- You dream of them often or feel their presence unexpectedly

- You compare new partners to them without wanting to

- You feel guilt, shame, or longing when trying to move on

- You struggle with sexual pleasure or clarity in future relationships

If this sounds familiar, you're not broken.
You're bonded. Energetically. Spiritually. Subconsciously.

How Soul Ties Are Formed During Sex:

- Orgasm opens the **auric field**, allowing energy to enter

- If ejaculation or vaginal fluids are exchanged, **DNA and vibration are shared**

- The womb acts as a **receptive vessel**, storing not just memory, but emotional resonance
- Eye contact, emotional vulnerability, or power imbalance during sex can **deepen the imprint**
- If the sexual experience involved **trauma, betrayal, or unresolved emotion**, the cord may fuse with pain

Why You Still Feel Them (Even If They're Long Gone)

Energetic cords operate outside of time. You may not love them anymore. You may have moved on logically. But if the tie was never acknowledged, healed, or released—your body still remembers.

Especially if:

- There was no closure
- The relationship ended in trauma or confusion
- The sexual bond was intense, addictive, or spiritually activating
- You gave parts of yourself away you didn't yet know how to protect

You are allowed to reclaim your energy. You are allowed to call your power back. You are allowed to feel, release, forgive, and let go. Because not everyone you've connected with deserves a lifelong seat in your spirit. You can bless the bond, learn the lesson, and then cut the cord.

And when you do, your soul breathes deeper. Your body softens. And your future stops being haunted by a past that was never meant to stay.

Karmic Entanglements & Spiritual Contracts

We Never Meant to Sign

Why Casual Sex Isn't Always Energetically Casual

We live in a world that promotes sexual liberation—but often forgets to teach **energetic accountability**.

Hookups. One-night stands. Casual flings. Moments of passion with strangers or situationships that go nowhere... You may walk away thinking the moment is over. But **your soul may still be entangled**. Because when you open your body, you don't just share pleasure.
You **sign spiritual contracts.**
You say:

> *"Come in."*
> *"I receive you."*
> *"I'll hold this."*

And sometimes what you hold is more than you bargained for.

Blood Bonds, Trauma Links, and Energetic Overlap

If sex involved **blood, emotional vulnerability, or orgasmic intensity**—and especially if the other person is also sexually active with others—you are now energetically connected to:

- **Them**
- **Their unresolved wounds**
- **Their other sexual partners**
- **The wounds of those partners**
- **Their ancestral grief, shame, or karmic patterns**

Energy doesn't stop with the person. It **webs out**—creating what you might call an **energetic orgy** of trauma, instability, addiction, spiritual confusion, or emotional imbalance.

This is why after certain sexual experiences, people report:

- Sudden mood swings, fatigue, or depression
- Heightened anxiety or fear that doesn't feel like their own
- Creative blocks or financial stagnation
- Obsessive thoughts about a partner they barely knew
- Nightmares, confusion, or increased spiritual interference
- Feeling "off" for weeks without explanation

These are not coincidences. These are karmic entanglements you didn't know you agreed to.

What Is a Spiritual Contract?

A spiritual contract is an **energetic agreement** that forms during deep exchange—like sex.

These contracts can bind you to:

- That person's energy, timeline, or karma
- Their thoughts, habits, and emotional states
- Their pain, insecurities, or spiritual baggage
- Unconscious soul-level pacts (like *"I'll carry your pain"* or *"I belong to you now"*)

These are often **not conscious**. But they are **real**. And if not revoked or cleared, they can affect your relationships, your manifestations, your clarity, and your peace.

Sex Is a Portal—and Portals Don't Close Themselves

When you enter sacred space with someone who is also entering it with others—who are also carrying unknown spiritual burdens—you open yourself to **collective energetic weight** that wasn't yours to hold.

It's not about shame. It's about **awareness.** Because liberation without boundaries is not empowerment—it's exposure. Not everyone deserves access to your body. Not everyone is meant to imprint your womb. And not every orgasm is worth the karmic invoice that comes after.

You are allowed to reclaim your body, your field, and your spirit. You are allowed to **rescind spiritual contracts** you never consciously signed. And you are absolutely allowed to return to yourself, whole, sovereign, and free.

Spiritual Contract Revocation Ritual

A Sacred Release of Energetic Ties You Never Consented To

> *This is for the lovers you didn't choose with clarity. The entanglements you didn't fully understand. The pain you absorbed, the portals you opened, and the contracts you didn't mean to sign.*

This ritual is your *return* to self.

It is your declaration that:

> *"I no longer carry what was never mine."*

Prepare Your Space: Sacred, Safe, Sovereign

- Find a quiet, undisturbed space
- Light a candle or incense
- Have a bowl of water, a journal, and something to safely burn paper with (fireproof dish or pot)
- Optional: play grounding or meditative music, use cleansing herbs (sage, rosemary, palo santo), or place crystals (black obsidian, selenite, or labradorite)

Take a few deep breaths. Place your hands over your womb or heart and whisper:

> *"I call all of my energy back to me now."*
> *"I am safe. I am sovereign. I am ready to release."*

Step 1: Write the Contracts

On small slips of paper or in your journal, write down:

- Names or initials of past sexual partners who still linger in your energy

- Experiences that involved trauma, confusion, coercion, or spiritual discomfort
- Words that were spoken or felt (e.g., "I belong to you," "I need you to feel whole")
- Any cords, sensations, or memories that still tether you emotionally or spiritually

You can begin each entry with:

> "I unintentionally created a spiritual contract with __________ when I __________."

or

> "I still feel energetically connected to __________ and I am ready to be free."

Step 2: Speak the Revocation Aloud

Hold your hands over your heart and repeat the following words out loud—slowly, intentionally, and from your soul:

> *"I now revoke any spiritual contracts, blood bonds, sexual ties, or energetic agreements*
> *that were formed without my full understanding, my clear consent, or my soul's alignment."*

> *"I break all cords of trauma, manipulation, codependency, lust, or soul-binding with any being who is not in alignment with my highest good."*

> *"I release their energy, their pain, their voice, their influence from my womb, my heart, and my field."*

> *"I call back every part of myself that was left behind, given away, or taken from me—whole and restored."*

> *"This portal is now closed. The contract is null and void. And so it is."*

Step 3: Burn the Papers

Carefully and safely burn the slips of paper in your fireproof dish or cauldron. As the flames rise, visualize the energetic cords dissolving in light. Feel the contracts disintegrating. Feel the space clearing.

Drop the ashes into the bowl of water.
Let the elements—fire and water—transmute what no longer belongs to you.

Step 4: Cleanse + Seal Your Energy

- Take a spiritual bath or shower after the ritual, using salt, rosemary, or lavender
- Imagine golden or violet light flooding your aura and womb space
- Anoint yourself with oil and say:

> *"I am whole. I am sacred. I am free."*

You are not bound by what you did in unconsciousness.
You are not sentenced to carry someone else's energy forever.
You are not defined by who entered your body—but by how you choose to return to yourself.

> *Your spirit is the final authority.*
> Your womb is yours again.
> Your future is unwritten.
> And your soul is now **unbound**.

Chapter 9

Marriage as Sanctity: The Ceremony of Evolving Love

From Sexual Union to Sacred Partnership

Marriage is not just a legal status. It is not a performance for society, or a checklist to impress others. Marriage is a **sacred ceremony**—one that continues long after the vows are spoken. It is a promise to not only love the person standing in front of you... but to **marry every version of them** yet to come.

Loving Through Evolution

As we grow, we **change**. Life introduces us to new truths, new wounds, new wisdom. We grieve, we heal, we stretch, we become. Your partner will not stay the same.
They will evolve—mentally, emotionally, spiritually.
Their desires may shift. Their fears may deepen. Their joys may expand.

True marriage says:

> *"I will meet you at every version of yourself—with grace, curiosity, and commitment."*

Sometimes, couples evolve **together**—in harmony, in sync. Sometimes, evolution **separates** what no longer aligns.

But that is why marriage is sacred: Because you are not just marrying *who they are now*— You are making a vow to love *who they will become.*

Why Casual Doesn't Deserve Covenant

You cannot give sacred privileges to temporary people.
You cannot keep offering your spirit to those who only

came for your body. Because sacred love requires **mutual responsibility**.

Yet today, many demand husband-and-wife treatment while offering **no real commitment**. They want your loyalty, your softness, your body, your womb—but they offer no leadership, no protection, no covering in return.

Discernment is not withholding love. It's honoring your **worthiness**. Because when you give too much to those not ready to carry it, you risk becoming **drained, disillusioned, and devalued**.

Marry the Man with a Provider Mindset

Not just a man with a wallet. But a man with **a heart, a vision, and spiritual maturity.**

The man who:

- Sees provision as *more than money*—as emotional presence, mental protection, and spiritual leadership
- Pays attention when you speak
- Shows up before you ask
- Creates a space where you can be soft, safe, and seen
- Doesn't just say "I love you," but proves it—**through action, sacrifice, and consistency**

A man with a provider mindset understands:

- Leadership is not dominance—it's responsibility
- Protection is not control—it's safety
- Presence is not perfection—it's **intentionality**

Be the Wife That Holds Space

While asking for covering, **become a covering** too. Hold space for his soul, not just his success. Be the one who sees behind the mask, who notices his silence, who knows when his spirit is tired but his pride won't admit it.

Create the kind of love that says:

> *"You don't have to be perfect here. You can be real."*

True partnership is mutual attunement. Learn him like you want him to learn you. Listen not just to his words—but to his energy. Be the place he **lays down his armor**. Because marriage is not just about romance. It's about refuge.

The Soul of the Vow

Don't marry potential. Marry **vision**. Marry **character**.
Marry **presence** and **principles**. Choose the partner who doesn't just impress—but **covers, sees, holds, and evolves** with you.

Because at the end of the day:

> *A man's value is not measured by his status—*
> *but by the safety, peace, and spiritual alignment he brings into your life.*

And a woman's love is not measured by what she gives away —but by how wisely she chooses where to **plant her sacred energy.**

Sex may bond. Love may spark. But **marriage is where souls intertwine in purpose.**

So choose the one who is prepared to carry every version of you. And be the one who honors every version of them— with truth, with tenderness, and with unshakable trust in the sanctity of evolving love.

Reflection Journal: Calling In Conscious Partnership

Envisioning Your Sacred Union

> *"If you don't know what your heart, body, and soul need—you'll keep settling for what looks good but feels wrong."*

This journal prompt is your space to dream, define, and align. Whether you're single, dating, divorced, or already married, this exercise helps you **tune in** to the kind of partnership your soul is calling in—or wants to reawaken.

Let your words flow. Be honest. Be unfiltered. Be bold.

1. What does *marriage as a sacred commitment* mean to me?

____
____
____
____

2. How do I want to feel in my future marriage or committed partnership?

☐ Safe
☐ Seen
☐ Desired
☐ Supported
☐ Respected
☐ Inspired
☐ Emotionally nurtured
☐ Spiritually aligned
☐ Free to evolve
☐ Other: ______________________

3. What traits do I desire in a spouse (or potential spouse)?

(Focus on character, values, energy—not just looks or status.)

- Emotionally available and honest

- Spiritually grounded or open to growth
- Protective and nurturing
- Consistent in word and action
- Emotionally intelligent
- Has a vision and values aligned with mine
- Makes me feel like home
- Listens deeply and communicates clearly
- Open to growth, healing, and evolving together
- Other:

4. Who do I need to become to align with this kind of love?

(Am I cultivating the traits I desire? Where can I grow or soften?)

5. What red flags or energetic mismatches have I ignored in the past—and what will I do differently now?

6. Soul Check: How will I know when I've found someone who's truly aligned with my spirit—not just my surface needs?

Final Manifestation Affirmation:

> *"I am aligned with the frequency of sacred partnership. I attract a love that honors my body, respects my heart, and awakens my soul. I choose wisely. I receive fully. I am ready for love that evolves with me."*

Part III

Embodiment, Boundaries & Rewriting Your Pleasure Story

Becoming the Guardian of Your Body, Desires, and Sacred Yes

> *"My body is no longer a battlefield for validation.*
> *It is a sanctuary. It is a sovereign space. And from this day forward—I decide what enters, what stays, and what sets me free."*
> —Dr. Deilen Michelle Villegas

You've explored the chemistry, the cords, the magic, and the meaning. You've remembered your body as a portal.
You've felt the depth of what intimacy truly is—spiritually, emotionally, and energetically.

Now comes the most important chapter of all:
Embodiment. Boundaries. Reclamation.

This is the part where you no longer just *know* the truth—
You start to **live it.** You speak your boundaries with clarity. You say yes to only what feeds your peace. You stop shrinking, performing, or settling. And you begin to **rewrite your story**—
not as someone who was broken...
but as someone who is **becoming whole.**

Welcome to Part III.

You are no longer seeking permission. You are no longer apologizing for your desires. You are no longer waiting for someone to choose you.

> *You are the altar.*
> *You are the offering.*

And you are the one you've been waiting for.

Chapter 10

Reclaiming Pleasure Without Performance

You Are Not a Stage—You Are a Sanctuary

From the time we are young, we are conditioned to **perform** our worth.

Smile even when you're tired. Be sexy, but not too sexual.
Make your partner feel good—even if you feel nothing. Be silent about your needs. Be desirable before you feel deserving.

And so many of us—especially women—learned to **confuse attention with intimacy**, performance with connection, and orgasm with obligation.

But true pleasure? True pleasure is not performative. It's **permission-based.** It's **body-led.** It's **truth-centered.**

Let Go of the Performance Mask

You don't have to:

- Moan louder than you feel
- Fake orgasm to avoid shame
- Say yes when your body is screaming no
- Hold your stomach in or keep your hair perfect
- Pretend to be turned on just to keep the peace

You are not here to entertain.
You are here to feel. To be seen. To be safe. To be sacred.

The more you perform, the more disconnected you become from your own sensuality. And eventually, you forget what **your pleasure** even feels like.

Reclaiming Starts With Unlearning

Reclaiming pleasure means:

- Listening to your body—without rushing or numbing
- Saying no when it's not aligned
- Saying yes only when it's *fully embodied*
- Releasing guilt, shame, and religious conditioning around desire
- Practicing intimacy that doesn't require climax to be valid
- Remembering that your pleasure is **your birthright**—not a prize for good behavior

What Real Pleasure Feels Like

Real pleasure is not a performance. It's a **presence**.

It can feel like:

- Warmth in your belly
- Softness in your chest
- Tingling down your thighs
- Laughter between kisses
- Stillness after breath

- Connection without pressure
- Slowness that feels safe
- A sacred "yes" that rises from deep within

If it's forced, it's not sacred.
If it's rushed, it's not embodied.
If it's performative, it's not pleasure.

Your body doesn't want a performance. It wants your **presence**. It wants to be trusted again. It wants to be heard when it whispers, "not yet," "not this," or "just a little slower."

This is your homecoming. This is where you stop pretending. This is where you **start remembering** what true, sovereign, sacred pleasure really feels like—on your terms, in your rhythm, in your body's truth.

Unlearning Shame and Religious Conditioning Around Desire

When Holiness and Wholeness No Longer Have to Conflict

Many of us were taught that our desire made us dangerous. That pleasure was sinful. That sexuality should be silenced, hidden, or offered only under very specific terms.

And so we learned to **disconnect from our bodies**. To separate the spiritual from the sensual. To view purity as disassociation. And desire as shame.

But what if **your body was never the problem**?
What if your desire was never dirty—but divine?

What Religion Often Gets Wrong

It's not desire that corrupts. It's **repression without**

understanding. It's moral shame without emotional context. It's the teaching that your body must be controlled, not loved. That your pleasure must be sacrificed, not explored with wisdom.

When we are taught that:

- Eve brought destruction through curiosity
- Virginity is purity, and sensuality is perversion
- Women must submit, men must dominate
- Pleasure outside of marriage is evil
- Masturbation, fantasy, or self-pleasure is wicked

We are not protecting the soul. We are **wounding the body**.

Shame, Control, and the Suppression of the Sacred Feminine

For centuries, organized religion has positioned itself as the gatekeeper of morality, purity, and power.

And in doing so—it has **stripped women of their sovereignty**. It has demonized desire. Punished sensuality.
Crucified the feminine.

Religion, in many forms, has taught women to see their bodies as burdens. Their sexuality as sinful. Their pleasure as something to be controlled, hidden, or shamed—unless it is for the benefit of a man.

The Double Standard of "Spiritual" Sexuality

In religious doctrine and patriarchal society alike:

- **Men are encouraged to explore, conquer, and pursue**.
- **Women are told to abstain, submit, and remain silent**.

This distortion glorifies the **objectification** of women but condemns their **liberation**.

If a woman expresses her desires, she's labeled "loose."
If she profits from her sensuality, she's labeled "a whore."
If she asserts her body as sacred, powerful, and intelligent—she's labeled "dangerous." This is not holiness.
This is **control disguised as divinity**.

The Legacy of Shame: A Tool of Suppression

Religious dogma has long used shame and guilt as tools of suppression:

- Teaching women that *modesty is moral*, but *pleasure is sinful.*
- Equating virginity with value.
- Silencing sexual trauma under the false belief that submission is spiritual.

Entire generations of women were raised to **fear their bodies**, to **deny their arousal**, and to **remain unspoken—even when in pain**.

But here's the truth:
The Divine Feminine is not silent. She is sensual. She is sacred. She is sovereign.

Psychology and Sociology of Religious Shame

Research shows that women raised in strict religious environments often experience:

- **Body shame and sexual repression**

- **Difficulty expressing sexual needs**
- **Increased risk of sexual dysfunction and trauma responses**
- **Low self-esteem tied to perceived "purity" or sexual history**

Study Insight:
A 2020 study in *Archives of Sexual Behavior* found that religious sexual shame correlates with **lower sexual satisfaction, higher guilt, and difficulty achieving orgasm**—especially among women.

"Sexual Shame and Religiosity in Emerging Adults" (Gulledge et al., 2020)

This Is a Spiritual Attack on the Feminine

Let us be clear—this isn't just a social pattern.
It's a **spiritual war** on the sacred feminine.

The systems of power have known for centuries that:

> *A woman who knows her body cannot be controlled.*
> *A woman who honors her pleasure cannot be manipulated.*
> *A woman who understands the divine power of her sensuality is unstoppable.*

So the feminine was cloaked in shame. Her magic called witchcraft. Her desire called sinful. Her voice called rebellious.

But *what if* the parts of you that were labeled "wrong"...
Were the very keys to your wholeness?

Reclaiming Your Sovereignty: From Shame to Sacred

Reclaiming your body is an act of resistance.

Reclaiming your pleasure is an act of liberation.
Reclaiming your truth is an act of divinity.

The path back to your power begins by:

- Naming the lies you were taught about yourself
- Unlearning what you were conditioned to believe was "unholy"
- Redefining intimacy as a source of healing, not shame

You were never meant to be small,
silenced, or submissive.

You were born to embody the full spectrum of
your sacredness—mind, body, and soul.

Religious trauma may have taught you
that your body is dangerous.

But in truth?

Your body is the altar. Your pleasure is the prayer.
And your sovereignty is the salvation.

What That Shame Becomes Later

- Disassociation during sex
- Fear of intimacy—even in marriage
- Belief that you are sinful for wanting touch or connection

- Difficulty asking for what you need in bed
- Numbness or guilt around orgasm
- Self-loathing around your curves, scent, or sensual expression
- A belief that your sexuality belongs to someone else—not you

> *You cannot hate your body into holiness. You cannot shame your desire into alignment. You cannot heal through suppression—you heal through integration.*

Reclaiming Sensuality as Sacred

Your body was made in divine image. Your pleasure is not an accident. It is **a portal to presence.** When you touch yourself with reverence— You are praying with your palms. When you allow yourself to receive—You are practicing trust in its purest form. When you explore your desire without shame—You are returning to **the original temple**: your body.

> *Your body is not a battleground. It is a cathedral.*
> *And pleasure is a hymn your soul still remembers.*

You can honor your spiritual values and still be fully in your body. You can love God and love sex. You can be both sacred and sensual, righteous and raw, pure and powerful. Because true divinity does not require your repression—It asks only for your presence.

And when you remember that your body is a home for the holy, You will never again confuse **shame with sanctity.**

How Religion Taught Women to Ignore Their Needs

The Silent Suffering Behind Sacred Scripts

For centuries, religion has shaped how women view their worth, their sexuality, and their right to receive.
And in the name of "purity" or "obedience," many were taught that:

- Sex is for men's release, not women's pleasure
- The only holy purpose for sex is procreation
- A woman's desire is sinful, distracting, or shameful
- Good wives don't ask—they endure
- Voicing your needs is rebellion, not righteousness

And so, **generation after generation**, women were told to:

- Lie still.
- Say nothing.
- Pretend it's enough.
- Let him finish.
- Bear the child.
- And be grateful.

Pleasure Was Painted as Sin—Except When It Wasn't His

Even in its most "basic" form, sex guarantees **one person pleasure**: the man. Whether love is present or not, whether foreplay exists or not, whether safety is offered or not—**he climaxes.**

But what about her? Where is her climax? Where is her fullness? Where is her voice? Her pleasure? Her preparation?

The narrative that sex is solely for procreation conveniently erases the **anatomical truth**:

> The clitoris exists for one purpose—**pleasure.** It has no reproductive function. Which means the female body was *designed by the divine* to enjoy intimacy.

So when someone tells you that pleasure is sinful, They're not upholding holiness. They're upholding **control.**

Patriarchal Sex: The Performance of Obedience

Let's be honest:
Most traditional teachings **center male orgasm** as the climax of sex and **neglect female anatomy entirely**.

Think about it:

- How many boys are taught where the clitoris is?
- How many men are taught to prioritize their partner's pleasure before their own?
- How many women are made to feel ashamed for guiding their partner—or exploring themselves to know what they like?

Religious dogma often reinforces:

- That women should not **masturbate** (and therefore, not know themselves)
- That women should not **ask** (and therefore, not communicate)
- That women should not **need** (and therefore, not receive)

- That sex is not about **connection and joy**, but about **duty and submission**

This is not divine order. This is **designated repression**—one that keeps women silent, disconnected, and performing roles they never consented to.

What Happens When Women Are Taught to Endure Instead of Enjoy

- They suppress their own pleasure
- They fake orgasms to preserve male ego
- They fear asking for what they need
- They internalize guilt for wanting more
- They feel "used" but cannot explain why
- They disconnect from their own anatomy
- They don't even know what's *possible*—because they were never shown

This leads to entire generations of women who:

- Have never had an orgasm
- Don't know what brings them joy
- Don't feel safe guiding a partner
- Stay silent in bed, afraid to offend
- Confuse suffering with submission

It's Time to Rewrite the Narrative

Pleasure is not rebellion. It's not dangerous. It's not unholy. It is **wisdom.** It is **intuition.** It is **your divine right.**

And when a man **loves you deeply**, when he is connected spiritually, emotionally, and mentally—he will **prioritize** you.

- He will want to **learn your body**
- He will **ask**, not assume
- He will view your pleasure as **a mirror of his maturity**, not a threat to his pride
- He will see your satisfaction as a **spiritual reward**, not a burden
- He will cover your body **with care, not control**

Because when a woman is cared for—**mind, body, and soul**—She opens. She follows. She flourishes. She gives from overflow, not obligation.

We are done apologizing for our needs. We are done pretending performance is pleasure. We are done believing silence is holy. Because when a woman's body is honored—When her pleasure is prioritized—When her voice is welcomed— She becomes not just a lover, but a force of sacred power. And that is how true intimacy begins—not with dominance, but with **devotion.**

The Language of Boundaries and How to Speak Your Needs Without Fear

Because Silence Will Never Lead You to Sacred Connection

You were not created to be silent. You were not made to endure discomfort in the name of keeping someone else comfortable.

You were not placed in this body to ignore it.

Boundaries are not selfish.

Boundaries are not mean.

Boundaries are **how you teach others to love you well.**

And yet for many of us, especially those who were taught to please, serve, submit, or endure...Speaking up can feel dangerous. Voicing needs can feel like rejection.
Saying "no" can feel like failure.

But here's the truth:

> *Every time you say yes when your body says no, you abandon yourself. Every time you suppress your needs to preserve someone else's comfort, you shrink. And every time you stay silent out of fear—you lose intimacy with the most important person in the room:* ***you.***

What Are Boundaries, Really?

Boundaries are **clear expressions of your needs, values, and limits**.

They are how you:

- Protect your body
- Honor your energy
- Define what is sacred
- Communicate what is welcome—and what is not
- Invite in healthy intimacy while protecting your peace

They are not ultimatums. They are invitations to **real connection**—rooted in **mutual respect and emotional**

maturity.

The Language of Boundaries Sounds Like:

- *"That doesn't feel good to me."*
- *"I'm not ready for that yet."*
- *"I need more foreplay before penetration."*
- *"Please don't speak to me like that."*
- *"I need some space to process this."*
- *"I feel disconnected. Can we slow down?"*
- *"I want to be touched like this instead."*
- *"I'm not okay with casual sex."*
- *"I need to feel emotionally safe before we're intimate."*

These are not demands—they are sacred declarations of truth. When spoken from the heart, they help filter what is real from what is performative. They attract partners who can honor your needs—and release those who cannot.

Why It's Hard (But Necessary)

Many of us struggle to speak our needs because:

- We were punished for voicing them in childhood
- We fear being "too much" or "too needy"
- We were taught to prioritize others over ourselves

- We were conditioned to believe that love must be earned by sacrifice

But love that **requires** your silence is not love. Intimacy that **demands** your submission is not sacred. Sex that **ignores** your needs is not connection—it's **extraction**.

Practice Makes Power

Start small. Practice saying no to things that drain you. Practice saying yes to what your body asks for.

Start with:

- Journaling your needs before you speak them
- Practicing phrases in the mirror
- Role-playing with a trusted friend
- Taking a breath before responding—giving yourself space to feel, then speak

And remind yourself daily:

> *"My truth is not a burden. It's a boundary. And the right people will honor it."*

Boundaries don't push love away. They call **real** love forward.

Because the people who are meant for you…Won't be offended by your voice. They'll be **grateful** for your clarity. And they'll meet you, not with resistance—but with **reverence.**

Chapter 11

Sacred Self-Touch and the Ritual of Coming Home to Your Body

Because Before You Can Be Touched by Another, You Must First Touch Yourself with Love

This is not just about masturbation. This is about **remembrance**.

Before your pleasure was distorted by shame...Before your body became a battlefield of opinions...Before your desire was silenced to soothe someone else's discomfort...You were whole. You were home. You were sacred.

And you still are.

Why Sacred Self-Touch Matters

Self-touch is not about replacing a partner. It's about **reuniting with yourself**.

It's how you:

- Reclaim autonomy over your body
- Rewire pleasure into safety instead of shame
- Explore what feels good without pressure to perform
- Create intimacy with your body on your own terms
- Begin healing from trauma, rejection, or disconnection
- Offer yourself the tenderness you always deserved

This is not about climax. This is about connection. It's about learning the landscape of your skin, your breath, your heartbeat, and your soul.

The Shift From Shame to Sacredness

Let's be clear: touching your own body is not dirty. It is not perverse. It is not forbidden fruit. It is an act of love. It is a prayer. It is a reclaiming.

So many of us were never taught how to love ourselves—only how to serve others. So when we touch ourselves, it's often with guilt, secrecy, or self-judgment.

Sacred self-touch asks:

- Can you slow down?
- Can you breathe?
- Can you be with your own body without rushing to an end goal?
- Can you explore without shame or comparison?
- Can you hold yourself the way you long to be held?

The Ritual of Sacred Self-Touch

This is not just an act—it is a **ceremony**.

Create a ritual space:

- Light a candle or incense
- Put on soft music
- Use oil or lotion that feels luxurious

- Begin fully clothed, simply placing your hands on your chest or belly
- Breathe deeply
- Whisper: *"I am safe in my body. I am home here."*

From there, allow your hands to explore:

- Your neck, shoulders, arms
- The inside of your thighs
- The soft of your belly
- The curve of your hips
- Your vulva or genitals—**only if and when it feels right**

You do not need to rush. You do not need to climax.
You are not here to achieve—You are here to **experience**.

Reclaiming the Divine Intelligence of Pleasure

Self-touch is not a sin. It is not shameful. It is not selfish.

It is sacred.

In spiritual circles, self-touch—whether sensual, emotional, or erotic—is a deeply personal practice of self-connection. It invites presence. It cultivates reverence. It reminds us that our bodies are **not just vessels for service or performance**—but portals of pleasure, healing, and knowing.

But beyond spiritual wellness, **science echoes the wisdom our ancestors always knew**:
Self-touch and masturbation are not only safe—they're

profoundly healthy.

Let's Talk Science: The Health Benefits of Masturbation

Far from the myths and taboos, masturbation is a **clinically supported** practice that contributes to:

- Stress relief
- Pain management
- Improved sleep
- Enhanced mood
- Hormonal regulation
- Boosted immunity
- Greater emotional clarity and self-awareness

Mental & Emotional Benefits

1. Reduces Stress and Anxiety
Masturbation increases dopamine and oxytocin—neurochemicals that reduce cortisol (the stress hormone).

> A 2019 study in *Frontiers in Psychology* showed that self-pleasure can significantly **lower stress and anxiety levels** through increased endorphin release.

2. Improves Sleep Quality
Orgasms stimulate a surge of prolactin and oxytocin, both of which support deep relaxation and quality sleep.

> According to the *Journal of Sleep Research (2021)*, individuals who climaxed (whether partnered or solo) reported **faster sleep onset** and **deeper rest**.

3. Emotional Regulation and Self-Soothing

Self-touch activates the **parasympathetic nervous system**, creating a calming effect. This makes it an excellent tool for emotional regulation, especially for those healing trauma or dealing with overstimulation.

Physical & Hormonal Health

4. Strengthens Pelvic Floor Muscles

Masturbation and orgasm cause involuntary contractions of the pelvic floor, promoting strength and blood flow in that region—especially beneficial for **postpartum women**, those with **menstrual pain**, or aging individuals experiencing **pelvic dysfunction**.

5. Reduces Menstrual Pain and Cramping

Orgasms increase circulation to the uterus and release pain-relieving hormones like endorphins.

> A 2020 survey published in *Sexual and Reproductive Health Matters* showed that 70% of women who used masturbation during menstruation reported **decreased cramping and improved mood**.

6. Supports Immune Function

Regular orgasm has been shown to **increase immunoglobulin A (IgA)** levels, enhancing immune defense.

> (*New England Journal of Medicine, 1999*) noted a **statistically significant increase** in IgA levels following regular solo sexual activity.

Self-Knowledge, Consent & Empowerment

7. Increases Sexual Self-Awareness

By engaging in self-touch, women develop a deeper understanding of:

- What they like

- Where they like it
- How to express it with confidence

This knowledge is **invaluable in consensual partnerships** and critical for identifying what does *not* feel safe or pleasurable.

8. Reduces Risky Sexual Behavior
Masturbation offers a **safe outlet for sexual expression**, reducing the likelihood of casual encounters that may compromise physical, emotional, or energetic well-being.

> A meta-analysis in *The Journal of Adolescent Health* found that individuals with healthy masturbation habits had **lower rates of STIs, sexual coercion, and unintended pregnancies.**

Self-Love Is Self-Regulation

When used intentionally, self-pleasure becomes:

- A grounding technique
- A ritual of presence
- A mirror of emotional truth
- A somatic pathway to release trauma and reclaim power

> "Self-touch is not just about climax—it's about connection."

It tells the nervous system: *You are safe here.*
It teaches the mind: *You are worthy of pleasure.*
It tells the body: *I see you. I honor you. I choose you.*

Research & Sources

- Brotto, L. A., & Smith, K. B. (2021). *Mindfulness and*

Sexual Well-Being: The Science of Mindful Sex. Journal of Sex Research.

- Levin, R. J. (2007). Sexual activity, health and well-being. *International Journal of Impotence Research.*
- Meston, C. M., & Buss, D. M. (2007). Why humans have sex. *Archives of Sexual Behavior.*
- Brody, S. (2006). The relative health benefits of different sexual activities. *Journal of Sexual Medicine.*
- Prause, N., & Pfaus, J. G. (2015). Viewing sexual stimuli associated with greater sexual responsiveness. *Journal of Sexual Medicine.*
- Mah, K., & Binik, Y. M. (2001). The nature of human orgasm: A critical review of major trends. *Clinical Psychology Review.*

You are not broken for wanting to touch yourself.
You are not shameful for needing connection.
You are not dirty for discovering what makes you bloom.

Self-touch is a revolution. It's medicine. It's your birthright.

Healing Through Self-Touch

For survivors of trauma or anyone disconnected from their body, self-touch can be **a portal back into safety.**
It's not easy. It may bring up grief, resistance, or numbness. That's okay. You're not doing it wrong. You're doing it with honesty.

Try saying:

> *"This part of me has carried so much."*
> *"You don't need to be perfect to be loved."*

"I forgive myself for abandoning you."
"I choose to return to you now."

You don't need to wait for someone else to wake up your body.
You don't need to wait to be chosen to be cherished.
Your hands are holy. Your touch is medicine.
Your body is not waiting for approval—it is waiting for **you.**

Let this be your homecoming.
Not to the mirror.
Not to perfection.
But to the soft, honest, sacred truth:
You are worthy of your own love.

Chapter 12

Body Worship, Mirror Work, and Sensual Confidence

Because Confidence Is Not Found—It's Reclaimed

This isn't about arrogance. This isn't about vanity.
This is about **devotion**. Devotion to your skin. Devotion to your curves, your scars, your softness, your strength.
Devotion to the vessel that has carried your soul through every heartbreak, every joy, every transformation.

> *Body worship is not sexualization. It is sanctification.*
> *It is saying: "I am holy. Even here. Especially here."*

Why We Struggle to See Ourselves Clearly

From the moment we could walk,
we were told:

To suck it in.
To cover it up.
To tone it down.
To smile through silence.
To compare, compete, conform.

We were told that our bodies were too much—
or not enough. Too round. Too brown. Too soft. Too loud.
Never quite right.

We inherited shame like second skin. Judgment like a mirror. And standards so high they broke our backs. So we learned to see ourselves through the lens of lack—Never through love. But that ends here. Because you cannot fully enjoy your sensuality while still at war with your reflection.

You cannot feel sacred in a body you've been taught to apologize for.
You cannot receive love in a body you've only learned to reject.
You cannot come home to yourself if you keep shrinking to fit someone else's vision.

This is your turning point. This is your mirror moment.
This is where you begin to **see clearly** again— Not with your eyes. But with your heart.

The Practice of Mirror Work

Mirror work is more than self-help fluff. It's a **soul retrieval** practice.

Here's how it begins:

1. **Stand or sit in front of a mirror**—naked or partially clothed, however you feel safe.

2. **Make eye contact with yourself.**

3. **Place one hand on your heart, one on your womb or belly.**

4. Say out loud:

   "I see you. I love you. I am coming back to you."

5. Move your hands over your body with tenderness.

   Trace the parts you've hated.
   Bless the parts you've hidden.
   Thank the parts that have held you through pain.

Let it be awkward. Let it be real. Let it be holy.

This is not a performance. This is a **return**.

What Is Body Worship?

Body worship is an intentional practice of **adoring your own body**—in word, in ritual, in touch, in presence.

Ways to practice body worship:

- Speak affirmations over each curve, roll, scar, and freckle
- Touch your skin with oil like you're anointing sacred ground
- Dance slowly in the mirror just to feel yourself move
- Take nude or sensual photos—not for attention, but for remembrance
- Look at your yoni, your chest, your thighs—not with critique, but with **curiosity and care**
- Ask yourself: *"If I believed I was divine, how would I treat this body today?"*

> *Worship doesn't start with lovers. It starts with you. Because no one can celebrate you louder than the shame you haven't yet silenced.*

Sensual Confidence Isn't Loud—It's Grounded

Sensual confidence isn't about volume. It's not about walking into a room and demanding attention. It's about **being so fully in your body** that your presence speaks before your mouth does.

It looks like:

- Wearing what *you* love
- Moving with intention, not apology
- Making eye contact with the mirror—and smiling
- Choosing softness as strength
- Receiving pleasure without guilt
- Walking like your hips remember they were sculpted from stardust

> *But sometimes, that confidence has been muted by years of social programming—so we must* ***call it back.***

Calling It Back Through Dance: Moving from Repression to Radiance

Indigenous cultures across the globe have long known:
Music and movement are medicine. They are how we pray. How we grieve. How we reconnect with joy. And how we shake off shame.

In many ancestral traditions, the **hips are the seat of sensuality, creativity, and emotional release.**
So when we move our hips, we don't just "look sexy." We **free stuck stories.** We **open locked portals.** We **shake loose generations of suppression.**

> *If trauma is stored in the body, then liberation must also be embodied.*

Mirror Dancing & Sacred Movement Rituals

You don't need choreography. You don't need an audience. All you need is **your breath, your rhythm, and your presence.**

Try this sensual dance practice:

1. **Create a safe, private space.** Light a candle. Put on music that makes you feel electric.

2. **Stand in front of a mirror—if it feels safe.** Gaze at yourself with no judgment. Just **witness**.

3. **Place one hand on your heart, the other on your lower belly.** Close your eyes. Breathe.

4. **Begin to sway—side to side, slow figure eights, gentle hip circles.** Let the rhythm rise from within, not from performance.

5. Let your arms float. Let your spine roll. Let your hands trace your body. Let your head fall back. Let yourself *feel.*

Move like you are seducing your own soul.
Dance like your hips are whispering your ancestors' prayers. Let the rhythm become a ritual.

Movements That Awaken Your Sensual Self

- **Hip Rolls** – Loosens stored tension and activates your sacral chakra

- **Pelvic Tilts & Circles** – Reconnects you to sexual energy and creative flow

- **Shoulder Rolls & Arm Undulations** – Opens your heart and softens your upper body

- **Spinal Waves** – Encourages freedom in your back, where we often carry emotional burden

- **Barefoot Grounding Steps** – Taps you into the earth, reminding your body it is safe to feel

You don't have to be a dancer to dance. You don't have to be perfect to be powerful. You don't have to perform to be sensual.

> *Your hips are altars.*
> *Your thighs hold stories.*
> *Your breath is a beat. Your body is a drum.*
> So let the rhythm **remember you.**
> Let the dance **resurrect you.**
> Let your sensual self **speak in motion.**

Because when you move from the inside out, you don't just become confident. You become **unapologetically, erotically, divinely free.**

Sensual Dance Playlist: Awaken, Move, Release, Remember

Choose music that stirs your soul, not just your hips. These songs are curated to guide you from **grounding to awakening**, into **exploration, liberation**, and finally, **embodied joy**.

Feel free to modify or create your own version based on mood, ancestry, or spiritual practice.

Grounding + Arrival

Slow, deep beats to help you connect to your breath and body.

- "Bad" – Wale, Tiara Thomas
- "Wading" – Jhene Aiko
- "To Love & Die" – Jhene Aiko
- "Yeah, I said It" – Rihanna
- "Good & Plenty" – Alex Isley, Masego, Jack Dine

Awakening the Fire

Start to stir your hips, shoulders, breath, and belly.

- "Call Me Every Day" – Chris Brown, Wizkid
- "Essence" – Wizkid ft. Tems
- "Toast" – Original Koffee
- "SoSo" – Kabusa Oriental Choir

Sensual Flow

Let yourself roll, sway, tease, and touch. This is where your inner goddess arrives.

- “Say Yes” – Floetry
- “Primetime” –Janelle Monae, Miguel
- “Goddess” – Banks
- “It Wont Stop” – Sevyn Streeter ft. Chris Brown

Liberation & Power Moves

Let it all go. Dance your truth, stomp your freedom, flip your hair.

- “Brown Skin Girl” – Beyoncé, Blue Ivy, Wizkid
- “Savage Remix” – Megan Thee Stallion ft. Beyoncé
- “Level Up” – Ciara
- “Freedom” – Beyoncé ft. Kendrick Lamar
- "Cold War" – Janelle Monae

Reflection + Return to Stillness

Breathe, ground, and honor the shift.

- “Stay" –Rihanna, Mikky Ekko
- "Hometown Glory” – Adele
- “Clocks" – Coldplay

- "All of Me" – John Legend

Tip: Create a playlist titled "My Sacred Sensual Self" so it's ready whenever your spirit needs to move.

Movement Reflection Journal Prompt

"What did my body teach me today?"

After your dance practice, sit in stillness and reflect:

1. What emotions surfaced as I moved? Joy? Shame? Power? Resistance?

2. Which parts of my body did I avoid? Which ones craved more attention?

3. Did I feel free? Awkward? Beautiful? Disconnected? Present?

4. What movements made me feel most alive or sexy?

5. Where did I feel stuck or stiff, and what might that say about what I'm holding?

6. What do I now understand about my sensuality that I didn't before I danced?

Final affirmation:

"I don't need permission to feel beautiful.
I give myself back to my rhythm.
I give myself back to my body.
And in doing so—I return to my power."

Chapter 13

The Power of Softness—Creating Safety Within to Deepen Connection With Others

Because the Way You Hold Yourself Becomes the Way Others Learn to Love You

We live in a world that rewards hard edges. Be strong. Be independent. Be unbothered. Be "that girl." Be "that guy".

But somewhere along the way, we forgot:
Softness is not the opposite of strength. Softness is how strength **feels** when it's rooted in peace. Softness is **safety made visible.**

> *You don't have to be hard to be powerful.*
> *You don't have to be guarded to be respected.*
> *You don't have to be cold to be protected.*
> You just have to be **rooted in your truth—and gentle with your own heart.**

Softness Begins With Self-Safety

We can't receive love fully if our bodies feel unsafe.
We can't relax into connection if our nervous system is still on alert.
We can't open to intimacy if our hearts are armored with fear, guilt, or old pain.

So the journey begins not with finding the right partner—but with **becoming a soft place to land for yourself.**

When you can hold space for:

- Your own tears without judgment

- Your triggers without shame
- Your boundaries without fear
- Your pleasure without guilt

You create an inner world where **your truth is welcome**. And when your truth is welcome—you attract people who can meet you there.

Softness Is the Language of Safe Love

In healthy connection, softness looks like:

- Being able to cry without apologizing
- Slowing down during sex to stay present
- Asking for what you need without guilt
- Allowing yourself to receive without feeling like you have to earn it
- Expressing vulnerability without feeling weak
- Trusting your "no" and honoring your "yes"
- Being fully seen—and still chosen

> *Softness isn't about always being calm—it's about being real.* It's about being **felt** instead of just being understood.

Practices to Cultivate Inner Softness

Reclaiming Your Right to Be Gentle With Yourself

Softness is not passivity—it is power wrapped in grace.
It's how we hold ourselves when the world forgets how.
And it's how we rewire the body to understand:

> *"You are safe now. You are home now. You are allowed to feel."*

Here are practices that support you in **coming home to softness**—physically, emotionally, and energetically.

1. Breathwork – Anchor Into Safety

Your breath is your first language of trust.
In moments of tension, fear, or shutdown, the breath becomes shallow or stops entirely. Returning to deep, conscious breathing tells the body:

> *"You are not in danger. You are safe to soften."*

Try this:

- Inhale for 4 counts
- Hold for 2
- Exhale for 6
- Repeat for 3–5 minutes
- As you breathe, gently say: *"I am here. I am safe. I am whole."*

This simple rhythm regulates your nervous system and prepares your body for softness, connection, and presence.

2. Gentle Self-Touch – Rewire the Relationship With Your Body

So many of us were taught to ignore, criticize, or perform with our bodies—never to *comfort* them. Gentle self-touch helps reintroduce the body to kindness, to consent, and to care.

Try this:

- Place one hand on your heart, one on your belly
- Breathe deeply
- Gently rub or press the space where emotion arises
- Say aloud:

"This part of me is safe now."
"I won't abandon you again."
"I choose softness over shame."

Let this be a ritual of reconnection—not correction.

3. Affirmations – Speak to Yourself With Reverence

Words create pathways in the brain. What you say consistently, you begin to believe. What you believe—you embody.

Choose words that invite your body to exhale and your spirit to rest.

Affirmations for softness:

- *"It is safe to soften."*
- *"I deserve tenderness—even from myself."*
- *"My softness is sacred, not weak."*
- *"I release the need to perform. I return to presence."*

Repeat them during mirror work, breathwork, or journaling

for maximum impact.

4. Mindful Movement – Let the Body Lead

Stillness isn't always possible after trauma—but movement can be medicine. Movement without force invites the body to reclaim joy, presence, and safety.

Try this:

- Slow, sensual stretching
- Hip circles or pelvic tilts
- Grounding walks barefoot in nature
- Restorative yoga or somatic dance
- Lying on the floor and swaying side to side, hands on your belly

Let movement be intuitive. No performance. No mirror. No pressure.

Just your body... returning to itself.

5. Emotional Check-Ins – Stay With Yourself

Many of us are quick to abandon ourselves emotionally—especially when big feelings arise.
Softness asks:

> *"What would it feel like to stay?"*

Practice this daily question:

> *"What do I need right now?"*

Follow with:

- Do I feel safe?

- Am I overwhelmed?
- Do I need space or connection?
- Can I be honest with myself right now?

Write the answers down. Let them guide your next soft action.

6. Letting People In – Be Seen, Slowly and Safely

Softness is also relational. Once you begin creating safety within, you must practice **inviting others into that safety**—gently and with discernment.

This can look like:

- Saying "I'm not okay" without shame
- Asking for a hug without fear of rejection
- Letting someone support you—even if you don't "need" it
- Naming your fear instead of hiding behind control
- Letting someone witness your process, not just your polished self

> *Connection deepens when you stop filtering your truth and start letting people see the softness under your strength.*

Softness is not something you find. It's something you **return to**—moment by moment, breath by breath, choice by choice.

And in that return, you begin to rebuild the relationship with the one person who matters most: **You.**

You are allowed to be soft and sacred. You are allowed to feel deeply and still be grounded. You are allowed to lay down your armor—without losing your power.

Because real love doesn't require performance.
It requires **presence**. And that begins when you stop protecting yourself from yourself.

> *You are not hard to love.*
> *You are not too much.*
> *You are a safe place waiting to be remembered—by you first, and then by the ones meant for you.*

Chapter 14

The Intimate Art of Being Seen—Emotional Availability in Love and Sex

Because the Greatest Turn-On Is Being Chosen for Who You Truly Are

We all want to be touched. But what we crave even more—is to be **seen**. To have someone look past the performance, the perfection, the persona—and still say:

> *"Yes. You. Just like that."*

But that kind of intimacy—the kind that melts armor and invites souls to meet—requires one thing many people struggle to give: **Emotional availability.**

Why So Many Can't Go There

Many people *want* deep intimacy...But their nervous system has been trained to protect, perform, or deflect.

They grew up in environments where:

- Vulnerability was punished
- Love was conditional
- Emotional needs were dismissed or ignored
- Survival required emotional suppression

So when it's time to be emotionally present, they:

- Pull away or go numb

- Intellectualize their feelings
- Become avoidant or reactive
- Use sex to replace emotional closeness
- Say "I'm fine" when they're hurting
- Crave connection but fear being seen

> *Emotional unavailability isn't a character flaw. It's a protection mechanism.* But protection is not connection. And love cannot bloom where the heart stays hidden.

What It Really Means to Be Emotionally Available

Your Body Can Be Present, But If Your Heart Is Guarded—You're Not Truly Here

Emotional availability is not about being perfect. It's not about always knowing the right words or having it all figured out.

It's about being *willing*. Willing to feel. Willing to stay. Willing to let someone see you beneath the layers.

In a world that teaches us to armor up, to dismiss feelings, to lead with logic and leave vulnerability behind—**emotional availability becomes an act of rebellion**... and of *sacred responsibility.*

Emotionally Available People:

Feel, even when it's messy

They don't run from sadness, anger, fear, or confusion. They lean in. They get curious. They sit with it. Because they know that feeling is not weakness—it's *awareness.*

And there's no intimacy without emotional awareness.

Name their emotions without shame

Instead of saying "I'm fine," they say, "I'm feeling overwhelmed." They don't guilt themselves for having big feelings—they respect those feelings as data. They give themselves (and others) permission to be *human.*

Hold space for their partner's feelings

They don't minimize, mock, or make it about them. They listen. They ask questions. They regulate their own responses. Because real love is *witnessing without needing to fix.*

Don't punish vulnerability

They don't weaponize someone's emotional honesty against them. They don't say "you're too sensitive" or withdraw love when truth is shared.
They understand that to be trusted with someone's vulnerability is a *privilege,* not an inconvenience.

Value presence over performance

They aren't distracted by the illusion of "perfect love" or polished appearances. They know real intimacy happens in sweatpants and silence, in awkward pauses and tearful truths.

> They aren't there just to be impressive—they're there to be *real.*

Sit with discomfort instead of fleeing

They don't ghost when conversations get hard. They don't distract themselves with sex, work, or withdrawal. They stay. They breathe. They return to the moment.

Because to be emotionally available means:

> *You are not just physically in the room.*

You are spiritually in the connection.
You are mentally engaged.
You are emotionally honest.
You are energetically ***open****.*

Emotional availability is not a personality trait. It's a practice. A choice. A daily invitation to show up **more honestly than you did yesterday.**

And it starts with you.

> If you want someone to hold your heart gently...
> You must first be brave enough to **hand it to them, whole.**

To be seen is not just to be looked at. It's to be **felt**, **witnessed**, **understood**, and **held**—in your truth, your tenderness, your terror, and your truth.

In a world obsessed with aesthetics and performance, **being emotionally available** is one of the most radical forms of intimacy. It says: "I am here—fully. I am not hiding. I am not performing. I am with you." But more importantly... **"I am with me."**

The Neurosomatic Science of Being Seen

Being seen—truly and safely—is not just emotional.
It's **biological**. It's **neurological**. It's **regulatory**.

Our bodies are wired for **connection** and **attunement** through a process called **co-regulation**—a core pillar of **Polyvagal Theory** (Porges, 2011). When someone is present, attuned, and emotionally available, they help regulate our nervous system through:

- **Facial cues** (soft eyes, gentle expressions)

- **Vocal prosody** (tone, cadence, inflection)

- **Touch** (warm, consensual contact)
- **Energy** (felt safety and groundedness)

This co-regulation leads to **ventral vagal activation**, the state where **trust, pleasure, and intimacy thrive.**
In contrast, emotional unavailability activates **sympathetic arousal** (fight/flight) or **dorsal vagal collapse** (shutdown/ disconnection), making love feel unsafe or performative.

> **To be emotionally available is a sign of a regulated nervous system.** It means your body isn't preparing for war—it's open to connection.

Why Emotional Availability Is Scarce—Especially in Men

Emotional availability is not just a personal trait. It is a **culturally shaped outcome**. And society, especially patriarchal systems, has done a devastating job of nurturing emotional availability in men.

From an early age, boys are told:

- "Don't cry."
- "Be a man."
- "Toughen up."
- "Stop acting like a girl."

These messages don't just create **emotional suppression**—they produce **nervous system dysregulation**. They **disconnect men from their bodies**, from their feelings, from their own internal compass.

They create men who:

- Shut down in conflict

- Can't name their feelings
- Are terrified of vulnerability
- Mistake dominance for intimacy
- Feel love, but cannot express it

> **"The emotionally unavailable man is often a deeply wounded boy still trying to survive a world that never made space for his softness."**

And this pain doesn't just stay inside—it leaks into relationships.

Emotional Availability in Women vs. Men

In Women:

- Tends to manifest as **emotional over-functioning**
- Hyper-attunement to others' needs (especially partners and children)
- May neglect their own boundaries and desires
- Often carry the **emotional labor** of the relationship

In Men:

- Often shows up as **emotional detachment** or avoidance
- May appear "logical," "calm," or "nonreactive" while emotionally shut down
- Difficulty expressing vulnerability unless in crisis

- May engage in **compartmentalization**—sexual availability without emotional intimacy

But beneath the surface, both are longing to be seen. Not just for the roles they play—but for the souls they carry.

The Psychological Need to Be Seen

Psychologist Carl Rogers famously said:
"When someone really hears you without passing judgment... it feels damn good."

This is because **emotional visibility** meets two of the deepest human needs:

- **Attachment security** (I am safe with you)
- **Authentic self-expression** (I can be all of me here)

When these needs are unmet, relationships begin to fracture—not from a lack of love, but from a lack of **emotional presence**.

The Cost of Emotional Unavailability

- **Chronic disconnection**
- **Mismatched intimacy needs**
- **Sex without presence**
- **Unresolved resentment**
- **Touch without attunement**
- **Loneliness in partnerships**

And here's the truth:

> You can't build soulful intimacy with someone who is emotionally absent—even if they are physically present. Presence without availability is a ghost wrapped in skin.

Rebuilding Emotional Availability

This is not about blaming. This is about **reclaiming**.

To be emotionally available, we must:

- Heal attachment wounds
- Learn nervous system regulation
- Develop somatic literacy (naming sensations and feelings)
- Practice honest, embodied communication
- **Create safety within** before expecting it from without

We must teach our sons and daughters that:

- Crying is not weak
- Expressing feelings is strength
- Vulnerability is not a threat—it's an invitation
- Intimacy is not just physical—it's **presence**, **attunement**, and **truth**

Research & References

- Porges, S. W. (2011). *The Polyvagal Theory:*

Neurophysiological Foundations of Emotions, Attachment, Communication, Self-regulation

- Siegel, D. J. (2010). *The Neurobiology of We: How Relationships, the Mind, and the Brain Interact to Shape Who We Are*

- Cozolino, L. (2006). *The Neuroscience of Human Relationships*

- Levine, P. A. (2010). *In an Unspoken Voice: How the Body Releases Trauma and Restores Goodness*

- Brené Brown (2015). *Daring Greatly*

- Van der Kolk, B. A. (2014). *The Body Keeps the Score*

To be seen… is to heal.
To be available… is to honor.

When love becomes a mirror, intimacy becomes transformation.
Your emotional presence is your greatest erotic gift.

Partner Reflection & Emotional Fluency Exercise

Build Emotional Safety. Practice Vulnerability. Deepen Connection.

> *"You can't ask someone to meet you deeply if you've never invited them into your truth."*

This guided reflection can be done **alone** (to prepare yourself for emotionally available love), or **with a partner** to enhance emotional intimacy and communication in real time.

Set the Space

- Choose a calm, distraction-free setting
- Sit facing each other, making eye contact (or journaling if solo)
- Light a candle or incense if desired
- Take three slow breaths together to ground your energy

Part 1: Emotional Fluency Check-In

Each partner takes a turn answering these prompts **honestly and without interruption**.

1. Today, I am feeling...

(Name at least 3 emotions—use both positive and challenging ones)
Examples: grateful, anxious, tender, overwhelmed, content, vulnerable

2. Right now, my body feels...

Where is there tension, ease, heaviness, or openness?

3. Something I'm afraid to say, but want you to know...

Say it slowly. Let the other just *listen*—no fixing, no reacting.

4. When I feel emotionally unsafe, I tend to...

Example: shut down, lash out, disconnect, over-explain

5. One thing that helps me feel safe again is...

Name a specific action, tone, or practice.

Part 2: Compassionate Witnessing (Partnered Version)

After each person shares, their partner says:

> "Thank you for trusting me with that. I see you. I'm still here. And I want to understand you better."

If doing this solo, place your hand on your heart and say it to your reflection or write it in your journal.

Part 3: Connection Prompts

Pick one or two of these to explore deeper:

- "When was the last time you felt truly seen by me?"
- "What's one thing you wish I asked you more often?"
- "What does emotional intimacy mean to you?"
- "What part of your emotional self do you still struggle to express?"
- "What helps you stay connected when we're triggered or distant?"

Safety Agreements (Optional)

To help hold this space with care, you may agree on:

- No interrupting

- No defending or correcting
- No making assumptions
- Time-outs are okay if overwhelmed
- Confidentiality is sacred
- All feelings are valid—even the uncomfortable ones

Integration Reflection (for Solo or Shared Journaling)

1. What did I learn about myself through this?
2. What surprised me or softened me?
3. What will I try to do differently going forward in love and communication?

Final Affirmation:

> *"I am capable of deep emotional intimacy. I welcome love that sees, hears, and honors me. I am safe to express. I am safe to receive. And I am worthy of being held with care."*

Emotional Availability in Sex

Sex without emotional availability is a performance.
It's two bodies colliding—while their souls remain strangers. But when presence enters the room...When vulnerability is welcome...When both people feel safe to express, pause, cry, laugh, or redirect...**Sex becomes sacred.**

You can:

- Share your desires without fear
- Slow down when something feels off

- Ask for what your body actually wants
- Feel your partner's heart—not just their hands
- Rest in the truth that *you're not being used—you're being met*

Because the best sex isn't always the wildest.
It's the kind where your body can exhale... and your soul can show up.

Signs You're Becoming More Emotionally Available

You May Not Feel "Healed"—But You're Becoming More Honest. More Present. More Real.

Emotional availability doesn't arrive all at once. It unfolds slowly, like petals opening to the sun after a long winter. Some days you'll still feel guarded. Other days, you'll surprise yourself with how open your heart has become. So how do you know you're growing? Look for the subtle shifts. The quiet bravery. The sacred unlearning.

You feel your feelings instead of avoiding them

You stop running from discomfort. Instead of numbing out, shutting down, or pretending, you pause and ask:

> *"What am I actually feeling right now?"*

You may still get overwhelmed—but now, you stay. You stay with the sadness. You sit with the anger. You welcome the joy without sabotaging it. You understand: **feeling is not weakness—it's evidence of life.**

You take responsibility for your triggers

Instead of blaming others for how you react, you take a breath and say:

> *"This reaction is mine to explore."*

You begin to notice your patterns:

- When you shut down during conflict
- When you avoid love that feels too good
- When old wounds echo into new relationships

And instead of shame, you meet yourself with **curiosity**.

> *"What does this part of me need right now?"*
> *"How can I respond, not just react?"*

You begin healing—not because you're trying to be perfect, but because you're learning to be present.

You speak your needs and boundaries clearly

You no longer whisper your truth. You don't shrink to be chosen. You don't expect others to read your mind.

You say:

- *"This is what I need to feel safe."*
- *"This is my limit."*
- *"This is what support looks like for me."*

And you say it with **love**—not apology.

> You no longer abandon yourself for the sake of peace. Because you've learned: *peace without truth isn't peace—it's performance.*

You let others witness you in moments of uncertainty

You don't wait until you have it all together. You let people in during the process. You say "I don't know" without shame. You cry in front of someone safe. You show your mess before the miracle.

You realize:

> *Being seen in your vulnerability is not dangerous—it's divine.*

You don't hide behind busyness, sarcasm, or sex

You used to use distractions to escape connection.
Work. Jokes. Sex. Silence. Now, you catch yourself. You pause. You get honest. You choose presence over performance. You no longer chase what numbs—you seek what nurtures.

> *You're not just in the room anymore. You're in the moment.*

You are learning to receive love—even when it feels scary

You let the compliments land. You stop questioning whether you're "too much" or "not enough." You soften when someone holds space for you. You don't run when someone shows up for real. Receiving love used to feel foreign. Now, it's becoming familiar. Not always easy. But possible.

> Because you finally believe: *I am worthy of being held, exactly as I am.*

Emotional availability isn't the absence of fear— It's the willingness to show up **with the fear... and stay anyway.**

It's crying without shame. It's being loved without resistance. It's being honest without apology. And if you're here—reading this, softening, showing up for yourself—

> *You're already on your way back to truth.*
> *To love. To connection that doesn't require you to hide.*

To be seen is one of the most sacred experiences on Earth.
But it begins when *you stop hiding from yourself.*

Because the truth is:

> *You don't have to be perfect to be loved.*
> *You just have to be present to be chosen.*

Emotional Availability Self-Assessment Worksheet

How Ready Are You to Be Seen, Held, and Loved?

> *"You can't build intimacy on avoidance.*
> *You can't receive love with a closed heart.*
> *You must be willing to meet yourself first."*

Use this worksheet to reflect on where you are on your emotional intimacy journey. Be honest. Be gentle. There is no "perfect score"—only information, insight, and opportunity.

Instructions:

For each statement below, rate how true it feels for you *right now* on a scale of 1 to 5:

1 = Not true at all
3 = Sometimes true / Working on it
5 = Very true / Embodied consistently

Statement	Rating (1–5)
I can name what I'm feeling without needing to justify it.	—
I give myself permission to feel the full spectrum of emotions.	—
I recognize my emotional patterns and know what triggers me.	—
I reflect on my reactions instead of blaming others.	—

I show myself compassion when I feel sad, angry, or overwhelmed.	___

Self-Awareness & Emotional LiteracyCommunication & Vulnerability

Statement	Rating (1–5)
I express my needs and desires clearly without guilt.	___
I feel safe saying "no" or setting boundaries.	___
I can talk about difficult emotions (fear, shame, loneliness) with someone I trust.	___
I let others see me—even when I'm not "put together."	___
I don't use humor, sex, or busyness to avoid deeper conversations.	___

Relational Presence & Receiving Love

Statement	Rating (1–5)
I can sit with emotional discomfort without immediately needing to fix or flee.	___
I make space for other people's emotions without judgment.	___
I don't punish or withdraw from people when they express vulnerability.	___

I am open to receiving love, support, compliments, or care—even when it feels unfamiliar. ___

I believe I am worthy of healthy, emotionally available love. ___

Reflection Questions

1. What patterns or habits am I proud of shifting so far?
2. Where do I still feel resistance or fear when it comes to being emotionally open?
3. What kind of love or connection am I truly ready for now?
4. What can I do to become a safer space for myself and those I love?

Integration Intention:

"I commit to meeting myself with honesty, compassion, and courage. I release shame, and I choose presence. I am no longer afraid to be seen."

Chapter 15

Rewriting Your Pleasure Story—From Pain and Performance to Presence and Power

Because Your Body Deserves a New Narrative

So many of us were taught that pleasure was something to hide, to fear, or to perform.

We learned that:

- Our bodies were made to be **controlled**, not celebrated.
- Our pleasure was **secondary**, if acknowledged at all.
- Intimacy was something to **endure**, not enjoy.
- And sex was a **transaction**, not a transformation.

We learned how to fake it before we ever learned how to feel it. We learned how to please others before we ever understood what pleased us.

But here's the truth:

> *Your body was never the problem.*
> *The stories you were told about it were.*

The Old Story: Pain, Performance, and Pleasing Others

Maybe your story sounds like:

- "I thought sex was supposed to hurt."
- "I don't even know what I like."

- "I've never had an orgasm."
- "I learned to moan on cue but feel nothing."
- "I disconnected to survive."
- "I stayed silent, because speaking up felt dangerous."

Or maybe your story has been masked in performative power:

- "I used sex to feel validated."
- "I knew how to turn them on, but not how to ask for what I needed."
- "I used seduction to distract from my sadness."
- "I looked confident, but I didn't feel connected."

These are not failures. These are **adaptive strategies for surviving in a world that taught us to fear our own bodies.** But survival isn't the goal anymore. **Freedom is.**

The New Story: Presence, Permission, and Power

Rewriting your pleasure story means:

- Saying *yes* because you *want to*—not because you're supposed to.
- Listening to your body instead of overriding it.
- Replacing performance with presence.
- Replacing silence with self-advocacy.

- Replacing shame with sacredness.

It means giving yourself:

- **Permission to explore.**
- **Permission to receive.**
- **Permission to pause.**
- **Permission to ask for more.**
- **Permission to say no.**

Because **pleasure isn't something you earn—it's something you reclaim.**

Presence Is the Path

Pleasure lives in the **now.** It doesn't live in how you look. It doesn't live in someone else's validation. It doesn't live in how much you know or how loud you sound.

> Pleasure is a breath. A slow touch. A laugh that escapes in the middle of something soft. A moment of full surrender—not to someone else, but to **your own sensation.**

You don't need to be "good" at sex. You don't need to be more flexible, more vocal, or more experienced. You just need to be **home** in your body. You just need to be **here**—not elsewhere in fear, not elsewhere in memory, not elsewhere in performance.

Just. Here.

You were never broken. You just needed someone to hand you the pen and say:

> *"Go ahead. You can write a different ending."*

And so you are. Page by page. Touch by touch. Breath by breath. This is not the story they gave you. This is the story **you chose**.

From Silence to Sensation—Discovering What Actually Feels Good to You

Because You Deserve to Know Your Body Beyond Obligation or Performance

For many of us, pleasure has been a performance. We were taught to moan on cue. To arch our back. To fake it 'til we make it. To prioritize their climax, not our comfort. To stay quiet, even when it hurt. To say nothing, even when it felt like nothing.

And somewhere in all of that—We forgot how to listen to our own bodies.Or maybe...We never learned how.

But you're here now.And it's not too late to begin again.
To reintroduce yourself to the body that has waited so patiently for your return.

> *This time, not with performance... but with presence.Not with pressure... but with permission.*
> *Not with silence... but with sensation.*

Why We Go Numb

Before we can rediscover what feels good, we must understand why we often *can't* feel at all:

- **Survival response:** If your body has associated touch with trauma, it will naturally disconnect from sensation to protect you.

- **Shame conditioning:** If you were raised to believe pleasure is wrong or dirty, your nervous system may shut down the moment it begins to awaken.

- **Hyperfocus on performance:** If you're too focused on *how you look*, you can't stay connected to *how you feel.*

- **Lack of communication:** If you've never been asked what you like, or felt safe to say it, your body may not have had the space to explore.

You are not broken. You are *disconnected.* And disconnection can be repaired—with gentleness, awareness, and time.

How to Reconnect With What Feels Good

This is not a race. There is no right answer.
This is an invitation to meet yourself in the present moment—no expectation, no performance, no shame.

Step 1: Create a Safe Container

- Choose a private, quiet space where you won't be interrupted

- Light a candle or incense to signal ritual

- Turn off distractions—this is a date with *you*

- Wear something that feels comfortable and sensual

Step 2: Engage in Sensory Mapping

- Touch different parts of your body—not just sexual areas

- Use various textures (hands, fabric, oils, feathers, etc.)

- Ask: *Does this feel soft? Does it feel irritating? Does it awaken anything?*

Try areas you've never given attention to:

- Behind your knees
- Your inner wrists
- Your scalp
- Your lower back
- The inside of your thighs
- The arch of your feet

Breathe slowly. Stay curious. No pressure.

Step 3: Journal What You Notice

Use these prompts:

- Where did I feel most sensation?
- What surprised me?
- Was there anywhere I felt numb, tense, or resistant?
- What kind of touch felt nurturing? Awakening? Neutral?

Give Yourself Permission To Say:

- "I don't like that."
- "I don't know yet."
- "That feels good."
- "I want more of that."

- "I need to pause."

You are learning a new language—*the language of your body.*And she is always speaking. The question is—**will you listen?**

You don't need to be an expert. You don't need a partner.
You don't need to *perform pleasure.* You just need a safe space, an open heart, and a willingness to feel again.

> *Because your pleasure is not a gift for someone else. It's a birthright you reclaim—for yourself first.*

Exploring Arousal Without Shame

Solo Touch as Sacred Self-Discovery

We can be honest:

Most of us weren't taught that touching ourselves was *sacred.* We were taught that it was "nasty," "wrong," "sinful," or "something boys do." For some, it was never mentioned. For others, it was *condemned.*

So we either:

- **Did it in secret**, cloaked in guilt
- **Didn't do it at all**, disconnected and disembodied
- Or we only explored our pleasure through someone else's hands, never our own

But here's the truth:

> *Your pleasure does not need to be earned.*
> *Your arousal is not dirty—it's divine.*
> *Your body is not shameful—it is sacred space.*
> And solo touch is not a betrayal of modesty—it is a

return to self.

Why Solo Touch Is Essential (Especially for Healing)

Solo touch is not just about orgasm. It is about **learning the language of your own body**—without performance, pressure, or external validation.

It teaches you to:

- Know what feels good *to you*
- Deepen your body awareness and pleasure vocabulary
- Build trust with your nervous system
- Reclaim consent and control
- Experience arousal without expectation
- Practice loving yourself from the inside out

And for trauma survivors, it allows healing at your own pace, in your own way—**without being watched, judged, or rushed.**

Releasing Shame: A New Narrative for Arousal

Arousal is a **spiritual frequency**.

It's not something to "turn off" or hide in fear. It's something to *honor, explore, and ground.*

Let's rewrite the script:

- *You are not wrong for having a body that feels.*
- *You are not shameful for craving connection.*
- *You are not impure for wanting to explore pleasure.*

- *You are not too much.*
- *You are not alone.*

Your arousal is not a problem. It's a **portal**—to intimacy, to healing, to divine remembrance.

Sacred Solo Touch Ritual: A Return to Self

Step 1: Prepare Your Space

- Dim the lights, light a candle, or play sensual music
- Use a towel, oil, or anything that supports softness and safety
- Speak aloud: *"I return to myself now. With reverence. With permission. With no shame."*

Step 2: Breathe and Set Intention

- Sit or lie down.
- Place your hands on your heart and womb
- Ask: *What does my body need today? Touch? Stillness? Exploration? Rest?*

Step 3: Explore Without Pressure

- Touch without a goal—no orgasm required
- Explore different strokes, rhythms, areas of the body
- Notice what feels good, what feels neutral, and what emotions arise

Step 4: Close in Gratitude

- Rest your hands on your body
- Say: *"Thank you, body. I will listen to you more. I will love you better."*
- Journal what you experienced, especially emotionally

Solo touch is not about replacing a partner. It's about returning to the one you've been seeking all along: **you.**

You are allowed to explore. You are allowed to feel.
You are allowed to know your own body—**not through shame, but through sacredness.**

> *Because when you stop making your arousal an apology, it becomes an offering.* And that offering is **holy.**

Pleasure Beyond Orgasm

How to Expand, Soften, and Stay Present Through Every Phase

We live in a culture obsessed with the finish line.
Orgasm is often treated like a goal, a prize, or proof that the act "worked." But pleasure—*real, sacred, embodied pleasure*—was never meant to be reduced to a single explosive moment. It is not just found at the peak.
It lives in the **entire journey**:

- The teasing spark of anticipation
- The warmth of skin against skin
- The inhale before the kiss

- The way your body arches without thinking
- The way your breath syncs to theirs
- The stillness after, where hearts keep whispering

Pleasure is not a race. It's a rhythm.
And orgasm is not the destination—it's just one stop along the way.

The Performance Trap: Chasing the Climax

Many people—especially those socialized as women—have been conditioned to:

- Pretend to climax to validate their partner
- Rush through arousal to get it "over with"
- Feel broken if they don't orgasm every time
- Believe that orgasm is the only "real" form of pleasure

But this creates anxiety, pressure, and disconnection from the *present moment.*

When you're performing, you're not feeling. When you're performing, you're not present.

True pleasure doesn't ask you to perform.
It asks you to participate.
Fully. Honestly. Gently.

Expanding Pleasure: What Happens When You Slow Down

When you stop rushing to orgasm, you begin to:

- Notice the micro-pleasures: the way your skin tingles, the beat of your own pulse, the energy building without needing release

- Experience pleasure as **waves**, not just spikes
- Feel more connected to your body, breath, and partner
- Allow deeper intimacy and emotional presence
- Reduce pressure and invite curiosity
- Heal trauma through slower, safer, self-honoring experiences

When you slow down, the body begins to trust.
And when the body trusts, pleasure flows more freely.

Staying Present Through Every Phase

Pleasure is not linear. It is cyclical, spiraling, and fluid.
Here are ways to stay present during each phase of your experience:

Anticipation:

- Breathe deeply. Feel the build. Welcome the longing.
- Say: *"I'm allowed to savor this moment."*

Arousal:

- Feel where the energy is building—don't rush it.
- Use your hands, breath, or sound to explore what feels *alive.*

Climax (if it happens):

- Allow yourself to surrender fully, without judgment
- Release expectations about how it should look or sound

Afterglow / Stillness:

- Stay connected. Hold yourself or your partner.
- Breathe together. Let your nervous system rest in safety.
- Say: *"This is still pleasure. I don't have to rush away from it."*

Pleasure isn't something to conquer. It's something to *commune with.* You are not broken if you don't climax.
You are not less worthy if your body responds differently each time.

> You are whole at every point in the process.
> You are worthy of slowness.
> You are worthy of softness.
> You are worthy of *presence over performance.*

Because when you stop chasing the peak—You realize you were standing on sacred ground all along.

Chapter 16

Releasing Guilt, Reclaiming Joy—Pleasure as a Revolutionary Act of Healing

Because Your Healing Doesn't Have to Be Quiet, Small, or Joyless

Somewhere along the way, many of us were taught that:

- **Joy should be earned** through struggle
- **Pleasure must be justified**
- **Happiness is suspicious** if it lasts too long
- And **desire is dangerous**—especially for women, especially for those raised in survival

So we learned to apologize for smiling too big. We made ourselves small in rooms where we wanted to be seen.
We swallowed our "too muchness" and buried our laughter in shame. We dimmed our joy because someone else was uncomfortable with it.

But here's the sacred truth:

> *Joy is not selfish. Pleasure is not sinful. Desire is not dangerous. They are your birthrights. Your medicine. Your rebellion.* And you do not need permission to feel good in your own body.

Guilt Is a Symptom of Conditioning, Not Morality

We often carry guilt not because we've done something wrong — but because we've internalized the belief that **we are wrong** for simply *feeling good.*

Especially if:

- You were raised in high-control religion
- You were parentified as a child and taught to serve, not receive
- You are a trauma survivor who only knows how to anticipate threat
- You were taught to associate self-worth with sacrifice and over-giving
- You are a woman, femme, BIPOC, queer, neurodivergent, or anyone raised in a system that benefits when you are disconnected from your power

Guilt says: *"Who do you think you are?"*
Healing says: *"I am someone who deserves to be well."*

Joy as a Healing Frequency

Pleasure doesn't just feel good—it heals. Joy doesn't just delight—it rewires.

When you allow yourself to feel joy in your body, you:

- Reclaim your nervous system from trauma loops
- Send safety signals to your brain
- Activate oxytocin and serotonin (the "connection" and "satisfaction" chemicals)
- Dismantle internalized shame
- Prove to yourself that *you can feel good and be safe at the*

same time

Joy is not the opposite of healing—it is the evidence that healing is working.
That you are not just surviving—you are beginning to **live** again.

Reclaiming Joy in Everyday Ritual

Revolutionary joy doesn't need a stage. It just needs space. You can reclaim joy in small, sacred ways every single day:

- **Move your body** to music that makes your hips remember they are magic
- **Wear lingerie, oils, or colors** that turn you on—even if no one sees you
- **Eat slowly and sensually**—with full presence
- **Say "no" without guilt**
- **Say "yes" to delight**—even when you're still healing
- **Laugh loudly. Cry fully. Moan without muting yourself.**
- **Celebrate your small wins** like they're sacred rituals—because they are

You don't have to "finish healing" to feel pleasure. Joy is not the reward. It's the **fuel.**

Let this chapter be a permission slip.

To be loud. To be soft. To be turned on. To be at peace.
To feel good in your skin—even if you were never taught how.

Because your joy **is not a luxury.** It is a **legacy**. A lineage of pleasure that your ancestors were denied—But you were born to reclaim.

> *You are the proof that healing can be beautiful.*
> *That softness can be strong. That pleasure can be sacred.* You are the revolution. And your joy is the uprising.

Epilogue:

Returning to the Sacred Self

You Were Never Lost—You Were Always Becoming

If you made it to this page, know this:

You are not the same woman, man, lover, or being who started this journey.

You've remembered.

You've reclaimed.

You've returned.

Not to a former version of you...But to the **truest** version of you. The one who never stopped pulsing beneath the shame, silence, and survival.

This book was never just about sex. It was about your **sovereignty.** It was about learning how to live and love in a body that feels like home. It was about **feeling again. Choosing again. Loving again.**

Your body is not a battleground. It is a temple. A compass. A portal. A poem. And now—you hold the pen.

Your Power Is in Your Presence

Pleasure is not a performance. Love is not a transaction. Intimacy is not something you earn by betraying yourself. You are allowed to ask for more. You are allowed to slow down. You are allowed to choose only what feels nourishing. You are allowed to say yes, no, not yet, or maybe later.

And most of all?

> *You are allowed to be loved without having to shrink, struggle, or prove a thing.*

You Are the Sacred Flame

May your body never again be a stranger to you. May you speak your needs with clarity, your boundaries with boldness, your desires with pride. May you rise from the ruins of shame, soft and sovereign.

And may every touch—yours or another's—be a prayer.

> *You are not too much.You are not broken. You are not behind.* You are sacred. You are sovereign. You are **home.** This is not the end.

This is your **beginning.**

Bonus Chapter:

Healing With Sovereignty:

Reclaiming the Mind, Body, and Spirit Through Ancestral Truth

A Holistic Call to Wholeness

In a medical culture that often treats symptoms while

ignoring root causes, I found myself drawn to the spaces in between, where science meets spirit, and where trauma is not just something that happens to the mind, but something that embeds itself in the body and echoes through the soul.

As a Board-Certified Holistic Health Practitioner, Clinical Mental Health Counselor, Trauma Specialist, and Certified Clinical Sexologist, my work exists at the intersection of clinical science, spiritual healing, and ancestral wisdom. I specialize in helping individuals navigate the layers of unresolved trauma, those silent, unseen wounds that often shape identity, health, relationships, and pleasure.

The Evolution of My Calling

My path as a healer didn't begin in a classroom. It began in the spaces between breath and breakdown, in the silent knowing I carried as a child, in the whispered prayers of my grandmothers, in the sacred rage I felt witnessing systems that failed to see us fully. My early work in Western medicine offered me clinical structure and scientific grounding. Still, I quickly saw the gaps, especially for women of color, for survivors of trauma, for those navigating grief and illness without language or support.

So, I expanded. I went back to the wisdom of the body. I listened to the ancestors. I studied plant medicine, breathwork, energy psychology, trauma therapy, and sacred sexuality. I remembered what so many of us are taught to forget: that healing is not linear, not sterile, not separate from spirit. Healing is cyclical. It is sacred. It is a return.

Over time, I became not just a practitioner but a bridge between science and spirit, between the clinic and the altar, between the past and the becoming. My clients don't come to me for surface-level solutions. They come when the soul is calling for liberation. And I meet them there, with truth, with tools, with tenderness, and with radical reverence for their process.

The Multidimensional Impact of Trauma

Trauma is not just an event; it is an experience that reprograms the nervous system, rewires perception, and often disconnects us from our own bodies. The Adverse Childhood Experiences (ACE) study showed us that early trauma can lead to long-term emotional, behavioral, and physiological consequences, including chronic disease, autoimmune disorders, and emotional dysregulation (Felitti et al., 1998).

But these findings, while groundbreaking, only scratch the surface. Trauma also impacts our sense of safety, our sexuality, our spirituality, and our ability to trust joy. In my practice, I've worked with clients who function highly in their careers, yet suffer silently in relationships, in their own skin, and within their womb space. This is where traditional talk therapy alone often falls short. Healing must be felt, not just spoken.

The Sacred Intersection: Sexual Health, Trauma, and Spirituality

One of the most misunderstood and often overlooked spaces in healing is the sacred intersection of sexual health, trauma, and spirituality. For many, sexuality is not simply a biological function or a pleasure-based experience; it is where our deepest wounds and highest potential live side by side.

As a Clinical Sexologist and trauma-informed practitioner, I've seen how trauma, especially sexual trauma, doesn't just affect the body or mind. It disconnects us from our very *source* of vitality: our sacral chakra.

For trauma survivors, especially those who have experienced sexual abuse, violation, or spiritual betrayal (through cheating and religious dogma), the reclamation of sexual agency can feel overwhelming. Yet, it is also one of the most profound doorways to liberation. I often say that the same place where the wound occurred can become the portal through which the healing happens.

In holistic medicine and energetic anatomy, the **sacral chakra** (Swadhisthana) is the energetic center located just below the navel. It governs our **sexuality, sensuality, pleasure, passion, creativity, emotional flow, and ability to experience joy**. When trauma is stored here, whether from abuse, cultural shame, betrayal, body dysmorphia, religious indoctrination, or reproductive grief, it often leads to disconnection from pleasure, repression of creativity, chronic guilt, numbness in intimacy, and even physical manifestations like pelvic pain, hormonal imbalance, or infertility.

In my practice, I guide clients to reconnect with their bodies through sacred sexuality, breathwork, and trauma-informed self-intimacy rituals. We explore shame, guilt, and fear not as obstacles to avoid but as invitations to transmute. By integrating body-based therapies with spiritual practice, including yoni healing, mirror work, ancestral rituals, and affirmation practices, my clients begin to see their sensuality as sacred, their bodies as temples, and their pleasure as prayer.

When we carry unresolved trauma in the sacral chakra, we often feel disconnected from our bodies and our desires. We may experience shame around intimacy, difficulty experiencing pleasure, a lack of motivation, or creative stagnation. We may find it hard to connect deeply with others or even to know what brings us joy. In this state, the energy of our sacral center becomes blocked, and that stagnation ripples into every area of our lives.

I've worked with women and femmes who are professionally accomplished, yet feel frozen in their bodies and disconnected in their relationships. They may struggle to find partners who truly value their feminine divinity. Many have been taught to disown their desire, silence their pleasure, or endure shame in exchange for survival. The result is stagnation, not just in sex, but in *life force*. They stop painting. Stop dancing. Stop dreaming. Stop desiring. They stop *receiving*.

Healing the sacral chakra is not just about sexuality. It is about **restoring life force** to the parts of ourselves that have been muted by pain. As we move stored trauma, awaken pleasure, and restore emotional flow, clients often rediscover their creative expression, reconnect with joy, and experience deeper satisfaction in both intimate and everyday relationships. This is the fertile ground from which passion, vision, and empowered decision-making grow.

This work isn't just about improving libido or enhancing relationships, although that often happens. It's about helping people come home to themselves. It's about giving them permission to feel again, to reclaim what was stolen, to write a new story where intimacy is rooted in safety, joy, and soul-deep connection.

This is where somatic practices, sacred sexuality, breathwork, and energetic clearing come together.

In my practice, I guide clients to move from:

- **Suppression to sensation**
- **Performance to presence**
- **Silence to self-expression**
- **Survival to embodied joy**

We explore sacred self-touch, yoni breathing, hip opening movements, and meditations designed to reconnect them with their inner fire. We call the womb space holy. We treat pleasure as a birthright. We affirm that desire is not dirty; it is divine.

Healing the sacral chakra also means facing and releasing what has been stored there for generations, colonial shame, patriarchal conditioning, sexual trauma, and even reproductive injustices. It's ancestral work. It's spiritual work. And it's often where clients first reclaim the *freedom to feel.*

When the sacral chakra is balanced, I've witnessed clients:

- Fall back in love with themselves
- Reignite passion in their relationships
- Launch long-dreamed-of creative projects
- Experience orgasms after years of numbness
- Begin feeling safe in their bodies, for the first time

This is not just about sex. It's about **sovereignty.**

When we reclaim the sacral space, we reclaim our power to create, to connect, and to choose pleasure as a sacred act of rebellion and remembrance.

Sexual Sovereignty & the Divine Portal of Manifestation

To reclaim sexual sovereignty is to awaken the divine blueprint encoded within us, the sacred energies of the **Divine Feminine and Divine Masculine** that live in harmony when we are fully embodied and spiritually aligned.

In my work, I often remind clients that our sexuality is not separate from our spirituality; it is one of its purest expressions. True sexual healing is not just about restoring intimacy or resolving trauma. It's about reuniting with the divine force within us that creates worlds.

The **Divine Feminine** within us is receptive, intuitive, fluid,

and creative.
The **Divine Masculine** is focused, active, structured, and purposeful.
When these energies are wounded, through trauma, conditioning, shame, or suppression, we struggle to manifest, to attract, to create.

What many people do not realize is that **our sexual energy is our creative energy.** It is the same frequency we use to build life, whether that's birthing a child, a business, a book, or a new version of ourselves. When our womb space is stagnant, we often find ourselves blocked, not just in pleasure, but in passion, productivity, and manifestation.

I have worked with countless women who were stuck, struggling to attract the love, abundance, or success they longed for. But beneath the surface, they were also disconnected from their bodies, burdened by shame, or had never truly experienced the sacred release of divine pleasure. And when that energy has nowhere to go... it turns inward, becoming anxiety, depression, self-doubt, or physical imbalance.

But when we begin to *reclaim our bodies as sacred*, something shifts.

Through intentional healing of the womb, the psyche, and the spirit, clients begin to experience a profound realignment. They stop seeking outside validation and begin embodying their desires. They activate the **energetic portal of the womb**, the sacred space not only for life, but for all creation.

When we awaken the womb space and connect to our sexual sovereignty:

- We co-create with the universe consciously
- We magnetize what we desire rather than chase it

- We speak with clarity, set boundaries with love, and choose partnerships that honor our worth

- We dream bigger, feel deeper, and move with divine confidence

This is the power of *sexual transmutation*, turning desire into reality, pain into purpose, and energy into manifestation. Pleasure becomes a prayer. Orgasm becomes activation. The body becomes a temple; not for performance, but for power.

In reclaiming our sexual sovereignty, we do not just heal, we *become whole*. We walk as the embodiment of sacred union: the Divine Feminine and Divine Masculine dancing in harmony within us. And from that place, we birth legacies, not just children, but revolutions.

The Holistic Framework I Use

My approach integrates:

- Neuropsychology & CBT/DBT techniques to reframe maladaptive thought loops
- Somatic Healing & Body Awareness to help clients release trauma stored in the nervous system
- Energy Work & Breathwork to restore spiritual vitality and emotional balance
- Herbal & Nutritional Therapy to support the body's biochemistry during healing
- Sexology & Sacred Sexuality to reconnect individuals with their pleasure and power
- Cultural & Ancestral Healing to honor lineage, identity, and unspoken grief

Case Study 1: Reclaiming the Body After Assault

"Tasha," a 36-year-old Black woman, came to me after years of sexual numbness and dissociation in her marriage. She had a history of childhood sexual abuse that had never been fully processed. Traditional therapy helped her cognitively understand her trauma, but her body remained frozen.

Using a combination of guided somatic experiencing, breathwork, and sacred self-touch practices, we slowly worked to reconnect her with her own sense of safety and sensuality. Over time, she was able to communicate her needs, experience pleasure without guilt, and reclaim her body as a sacred vessel, not a site of pain.

Case Study 2: Trauma in the Womb Space

"Lena," a 28-year-old Afro-Latina, came seeking help with PCOS, irregular cycles, and unexplained pelvic pain. Though initially seeking support for hormonal balance, our sessions uncovered years of inherited trauma from her mother's experience of forced sterilization.

We incorporated womb journaling, yoni steaming, trauma-informed herbal blends, and grief rituals to help Lena reclaim her womb as a center of wisdom, not just pathology. Today, she reports more regular cycles, increased self-trust, and a deepened sense of spiritual embodiment.

Case Study 3: Spiritual Dissonance and Soul Retrieval

"Marcus," a 42-year-old Afro-Caribbean man, came to me in the aftermath of a life-altering accident. Though his physical injuries had healed, he experienced persistent emotional numbness, panic attacks, and an overwhelming fear of abandonment. What we uncovered in session was unresolved grief linked to both ancestral disconnection and spiritual suppression growing up.

Through a series of soul retrieval ceremonies, breathwork, ancestral altar building, and psychoeducation around intergenerational trauma, Marcus began reconnecting with his cultural roots and rediscovering his own spiritual gifts.

The panic attacks diminished. The dreams returned. And with that, a powerful remembering took place.

The Research That Validates This Work

- Bessel van der Kolk's *The Body Keeps the Score* (2014) confirms that trauma is stored in the body, not just the brain.
- Peter Levine's *Somatic Experiencing* method has demonstrated success in helping clients release chronic fight/flight/freeze patterns through bodily sensation tracking.
- Dr. Gabor Maté's work in trauma and addiction highlights how emotional repression becomes dis-ease over time.
- Emerging studies in epigenetics show how trauma can be passed down generationally through changes in gene expression (Yehuda et al., 2016).

All of these affirm what many of us in holistic spaces have known through lived experience: healing must happen on all levels, mental, physical, spiritual, and ancestral.

Why This Work Matters Now

We are living in a time of collective awakening, and collective distortion.

Many are beginning to recognize that true wellness can no longer be separated from culture, spirituality, embodiment, or ancestry. Yet at the same time, we are witnessing the effects of centuries of **colonialism, cultural erasure, and spiritual theft**, especially as it relates to the body, sexuality, and sacred feminine energy.

Colonization did not just steal land. It **stole rituals**, erased lineages, demonized the womb, and replaced sacred sexuality with shame. It took what was once holy, the dance of sensuality, the wisdom of women, the power of creation, and turned it into something to be exploited, controlled, or suppressed.

What we see today as **hypersexuality**, particularly in media, entertainment, and cultural narratives, is not a reflection of freedom; it is a symptom of spiritual **disconnection and overcompensation.** It is the result of an **overactive sacral chakra**: desire without devotion, stimulation without soul. Women's bodies are commodified. Sexual energy is extracted, not exchanged.

Sex has been reduced from a sacred ceremony to a **performance** and **consumption**. The womb, once honored as the cosmic portal of life and manifestation, has been stripped of reverence and turned into an object of conquest. And in this dynamic, **women, particularly Black, Latin, and Indigenous women, have been cast as sources of energy for others, while being depleted of the ability to use that energy for their own creation and healing.**

In this current climate, **sexual sovereignty is revolutionary.** To reclaim our sensuality, our sacred boundaries, our divine pleasure, not for performance, but for *power*, is a return to ancient knowing. A remembrance of what was almost forgotten.

When we restore the **true meaning of sexuality**, we remember that it is:

- A divine portal
- A tool for manifestation
- A mirror of our internal wholeness
- A vehicle for ancestral healing
- A sacred contract, not a casual transaction

And when women remember this, when we return to the womb as the seat of wisdom and the source of our creative magic, we disrupt every system that has ever profited from our disempowerment.

This work matters now because we are restoring the balance. We are reactivating the truth of our bodies, our lineages, and our pleasure. We are shifting from survival to sovereignty. From silence to sacred speech. From depletion to divine embodiment.

This is not just about healing trauma. It is about **reclaiming our legacy.** And when we do, we become architects of a new paradigm, one rooted in wholeness, remembrance, and liberation for generations to come.

Ritual for Reclaiming Sacred Sensuality

A Ceremony of Remembrance, Embodiment, and Divine Feminine Activation

This ritual is not about performance. It is not about being desirable to others. It is about *becoming desirable to your own soul again.* It is a sacred reclamation of your sensual energy, your body's wisdom, and your right to feel, fully, fiercely, and freely.

What You'll Need:

- A quiet, sacred space
- A mirror (handheld or full-length)
- A journal and pen
- Optional: a red or orange candle, rose or ylang-ylang essential oil, music that moves you, a soft scarf or cloth, fresh flowers
- A small bowl of water and a pinch of salt (for cleansing)

Step 1: Cleanse & Ground

Begin by taking a few deep breaths.
Place your hands on your lower belly, the seat of your sacral chakra, and say aloud:

> *"I call all parts of myself back to me now.*
> *I call back the power I gave away, the pleasure I denied, the voice I silenced.*
> *I return to the body. I return to the breath. I return to myself."*

Gently wash your hands or face with the salt water.

If you have a candle, light it. If you have oil, anoint your womb space or inner thighs with care.

Step 2: Sensual Self-Connection

Stand or sit in front of your mirror. Look at yourself, not through the lens of critique, but through the eyes of the Divine. This is your temple. This is your vessel. This is your home.

Let your hands softly explore your body, your arms, your chest, your hips, with reverence. Breathe into areas where you feel tension or numbness. Let your body move intuitively, sway, stretch, dance, or stillness. Let it speak.

Repeat aloud or silently:

> *"My body is sacred. My sensuality is holy. My pleasure is power. I am not here to be consumed. I am here to create."*

Step 3: Womb Whispering & Journaling

Place both hands over your womb. Close your eyes and breathe deeply into this space.

Ask:

- *What do you want me to remember?*
- *What do you need to feel safe, open, and alive?*
- *What am I ready to release?*
- *What am I ready to receive?*

Write down what you hear, feel, or intuit in your journal, no filters.

Step 4: Affirm & Activate

Speak this affirmation aloud (or write your own variation):

> *"From this moment forward, I walk in sacred sensuality. I honor my body as a temple, my womb as a portal, and my pleasure as a prayer. I will no longer silence my desire or sacrifice my energy.*
> *I reclaim my power to feel, to create, to receive.*
> *I am sensual. I am sovereign. I am sacred."*

Close the ritual by thanking your body, your womb, your ancestors, and your inner Divine. Blow out the candle. Place the journal on your altar or under your pillow.

Optional Integration Practice (Over 7 Days):

- Wear colors of the sacral chakra (orange, coral, deep red)
- Do hip circles or womb yoga each morning
- Say your affirmation while placing your hand over your womb
- Eat nourishing, juicy foods (mangoes, coconut, berries)
- Avoid content that objectifies your body, replace it with music, art, or poetry that exalts it

This is how we come home. Not through someone else's gaze, but through our own sacred remembering. You do not need to be fixed; you are returning to your fullness.

And from this fullness, you will manifest worlds.

Practitioner's Afterword:

A Love Letter to the Body, the Spirit, and the Sacred Truths Within

To the Seeker, the Lover, the Survivor, and the Sovereign— You have arrived at the end of these pages, but the beginning of something far more important: a return to yourself.

As a Board-Certified Holistic Health Practitioner, Clinical Mental Health Counselor, and Clinical Sexologist, I have walked beside clients as they battled shame, reclaimed pleasure, and rewrote the stories embedded in their skin. I have witnessed the ways trauma tightens the body, numbs desire, and silences the soul. And I have also seen the miracles that occur when a person feels *safe enough* to feel again.

This book was never just about sex.
It was about *sacred remembrance*. About reclaiming the wisdom your body has always known. About transforming the disconnection caused by societal programming, religious shame, and cultural trauma—and replacing it with intimacy, sovereignty, and embodied truth.

The Science Is Clear

Neuroscience tells us that intimacy is not a frivolous luxury —it's a biological need. The nervous system regulates itself through touch, presence, and connection. Oxytocin calms our inner storms. Clitoral stimulation can relieve migraines. Masturbation helps balance cortisol and builds a stronger relationship with self. Sexual wellness *is* health.

Trauma-informed care teaches us that safety, choice, and attunement are the cornerstones of healing. And sexual healing is no exception. This is why many struggle with pleasure: not because they're broken—but because they've never been held in a way that honors their nervous system.

The Spirit Never Forgot

Our ancestors revered the body as sacred. Blood as holy. Pleasure as a portal to the divine.

Yet colonization, patriarchal religions, and capitalist agendas taught us to fear our flesh. To demonize desire. To glorify suppression while criminalizing expression.

But you, dear reader, are part of the rising. You are the embodiment of divine memory reawakening. And every time you choose intimacy that is conscious, consensual, and connected—you participate in the revolution of remembrance.

To the Practitioner Reading This

May you use this book as both mirror and map. Let it deepen the way you serve your clients, hold space, and speak of pleasure without shame. Normalize conversations about vaginal pain, erectile health, sacred boundaries, and sexual trauma with the same reverence you give any other system in the body.

Teach your clients that knowing their bodies is not taboo—it's *vital*. That they can redefine sex, rewrite the script, and reclaim their voice—on their terms.

To the Reader on the Path

There is no "right" way to be intimate—only *your* way.

Whether you're learning to receive love again after betrayal, building a connection with your womb, deconstructing purity culture, navigating your own gendered experience of pleasure, or simply seeking to know your body better—know this:

You are not too much. You are not too late. You are not broken. You are *becoming*.

Thank you for journeying through these pages with courage and curiosity. The flesh is sacred—not because it is flawless —but because it holds the imprints of your resilience, your longing, your wisdom, and your divinity.

Stay devoted to your body's truth. Stay open to love that honors all of you. And most importantly—never again apologize for your pleasure.

With reverence, fire, and fierce tenderness,
Dr. Deilen Michelle Villegas, Ph.D., DNM, HHP, ABS, CMHC
Holistic Trauma Specialist | Clinical Sexologist | Keeper of Sacred Flesh

Resources for Continued Healing, Exploration, and Empowerment

Curated with care to support your journey through sexuality, intimacy, trauma healing, and spiritual embodiment.

Books on Sex, Pleasure, and Embodiment

- *Come As You Are* by Emily Nagoski
- *The Body Is Not an Apology* by Sonya Renee Taylor
- *Women's Anatomy of Arousal* by Sheri Winston
- *The Erotic Mind* by Jack Morin, PhD
- *Healing Sex* by Staci Haines
- *Pleasure Activism* by adrienne maree brown
- *Tantra Illuminated* by Christopher Wallis
- *Sexual Healing Journey* by Wendy Maltz

Somatic Healing + Trauma Recovery

- *The Body Keeps the Score* by Dr. Bessel van der Kolk
- *Waking the Tiger* by Peter Levine
- *The Wisdom of Trauma* by Dr. Gabor Maté (film & website)

- *Polyvagal Theory in Therapy* by Deb Dana
- *Anchored* by Deb Dana
- *My Grandmother's Hands* by Resmaa Menakem

Spiritual & Energetic Tools

- *The Sevenfold Journey* by Anodea Judith & Selene Vega
- *Energetic Boundaries* by Cyndi Dale
- *Chakra Healing* by Margarita Alcantara
- *Sacred Woman* by Queen Afua
- *Jambalaya* by Luisah Teish
- *The Sophia Code* by Kaia Ra

Holistic & Culturally Centered Healing

- *Decolonizing Wellness* by Dalia Kinsey
- *The Inner Work of Racial Justice* by Rhonda V. Magee
- *Sister Outsider* by Audre Lorde
- *The Four Agreements* by Don Miguel Ruiz
- *All About Love* by bell hooks
- *Sensual Faith* by Lyvonne Briggs

Websites & Online Resources

- Trauma Research Foundation

- The Center for Mind-Body Medicine
- The Gottman Institute (for relationship tools)
- Afrosexology
- Shadeen Francis – Licensed sex therapist & educator
- The Shamanic Goddess – Dr. Deilen Michelle Villegas' holistic wellness offerings

Podcasts & Guided Practices

- *Sex with Emily* – Dr. Emily Morse
- *Sensual Self Podcast* – Ev'Yan Whitney
- *Dear Sex* – Hosted by Stephanie Esters
- *Therapy Chat* – Laura Reagan, LCSW
- *Unlocking Us* – Brené Brown
- *Medicine Stories* – Amber Magnolia Hill

Ritual Tools & Healing Modalities to Explore

- Womb steaming (yoni steaming)
- Tantric breathwork + eye gazing
- Somatic experiencing therapy
- Reiki + energy healing

- Sacred dance/movement meditation
- Sound healing + frequency therapy
- Ancestral reverence + altar work
- Sensual journaling + mirror work
- Oracle decks for embodiment guidance

Hotlines & Immediate Support

- **RAINN (Rape, Abuse & Incest National Network)**
  800.656.HOPE | www.rainn.org
- **Crisis Text Line**
  Text HOME to 741741 | www.crisistextline.org
- **National Suicide Prevention Lifeline**
  988 (U.S.) | 988lifeline.org
- **Therapy for Black Girls**
  www.therapyforblackgirls.com
- **Inclusive Therapists**
  www.inclusivetherapists.com

Appendix A: Partner's Guide to Touching Her Sacredly

A Sensual & Somatic Exploration Worksheet for Lovers

This worksheet is designed to help partners create **intentional, reverent, and embodied intimacy**—moving beyond performance into deep presence. Touch is not just physical—it's energetic, emotional, and spiritual. This guide invites you to slow down, tune in, and touch with devotion, not assumption.

Step 1: Set the Sacred Container

Before any touch begins:

- ❑ **Have we created a space of emotional safety?** (Discussed boundaries, fears, or intentions beforehand)

- ❑ **Have we cleared distractions?** (Phones off, dim lighting, music that soothes the nervous system)

- ❑ **Have we asked for consent—not just for sex, but for sacred touch?**

  Tip: Say, "I want to explore your body with care and presence. Is now a good time for that?"

Step 2: Connect Before You Caress

Practice this grounding ritual:

- Sit face to face.

- Breathe together for 1–2 minutes.

- Gaze into each other's eyes without speaking.

- Place your hand over each other's hearts.
- Whisper a *one-word intention* for this experience.

Step 3: Touch Types & Her Somatic Cues

Check in on how she receives **different forms of touch**. Use a scale of:
1 = dislike
5 = love it
B = causes overthinking
Y = feels arousing
G = feels relaxing
N = don't do this without emotional safety

Type of Touch	1–5	Notes/Body Response
Light fingertips		
Palm of hand		
Firm pressure (massage)		
Circular motions		
Tapping or drumming		
Tongue / licking		
Breath on skin		
Hair / scalp stroking		
Inner thigh tracing		
Sacred spot exploration		

Step 4: Ask, Listen, Adjust

Use these prompts to spark conversation:

- "Where on your body craves touch today?"
- "What kind of energy do you want this touch to carry—love, worship, play, grounding?"
- "What makes you feel most *invited*, not just taken?"
- "Are there places you want me to avoid today?"
- "What's one thing I did last time that you want more of?"

> *Remember:* Arousal is **not linear**. Presence is more important than performance.

Step 5: Energetic Intimacy Check-In

After the experience, reflect together:

- "Did anything come up emotionally?"
- "Did any part of your body tighten, freeze, or pull away?"
- "What helped you feel safe and surrendered?"
- "How can we deepen this experience next time?"

Seal the Energy

At the end, hold her.
Say **thank you**.
Look her in the eyes and affirm:

> *"Your body is sacred. Thank you for trusting me."*

Then breathe together in stillness. Let the energy settle. Let the connection anchor.

Appendix B: Before You Enter Me, Enter My Energy

A Partner Communication Guide for Sacred Intimacy

Intimacy is not just physical—it is energetic, emotional, and spiritual. This guide offers couples a powerful tool to pause, connect, and communicate before their bodies come together. It is an invitation to **honor each other's energy before engaging in erotic exchange.**

When you **touch her body**, you are **entering her field.**
When you **penetrate her womb**, you are **entering her temple**.
When you **share your essence**, you are **leaving an imprint.**

This is a sacred exchange. Prepare for it accordingly.

Step 1: Grounding Before Intimacy

Ritual Before Romance
Take 3–5 minutes to ground your energies individually. Then face one another and ask:

- ❑ "How are you feeling today emotionally, mentally, and energetically?"

- ❑ "Is there anything weighing on your heart that we should acknowledge before we connect?"

- ❑ "Are you feeling open, neutral, or closed in your energy right now?"

Note: This is not meant to be a therapy session—but a respectful energetic check-in that promotes safety and presence.

Step 2: Energy Consent

Ask for permission on the energetic level:

- “May I enter your energy before I enter your body?”
- “Is there any part of your energy that’s calling for tenderness, patience, or slowness today?”
- “Would you like touch, presence, eye contact, or stillness first?”
- “Can I offer you something before I take anything from this experience?”

Energy Consent is real. Honor the yes. Honor the no. Honor the maybe.

Step 3: Emotional Availability Check

Ask and reflect:

- “Do you feel emotionally safe with me right now?”
- “Is there anything I did recently—intentionally or unintentionally—that may have disrupted your trust?”
- “What would help you feel seen, supported, and cherished in this moment?”
- “What do you want from this connection—release, play, healing, affirmation, or presence?”

Step 4: Speak the Love Languages of Energy

Understand that touch isn’t always the first door. Use these prompts:

Love Language	Energy Translation	Partner Reflection

Words of Affirmation	"I honor your body as sacred." "You are safe with me."
Acts of Service	"Can I help you relax or ground before we begin?"
Receiving Gifts	"I brought something to nurture your energy —incense, oils, tea."
Quality Time	"Let's sit in silence together before we engage."
Physical Touch	"May I hold your hand while you breathe and settle?"

Step 5: Energetic Boundaries & Agreements

Before intimacy, ask:

- "Are there any emotional or physical limits tonight?"
- "If anything feels off or overwhelming, how should we communicate that?"
- "What is your sacred safe word or signal?"
- "What energy are we calling in tonight—playful, devotional, healing, sensual, wild?"

When partners co-create the energy of intimacy, it deepens trust, embodiment, and pleasure.

Step 6: Post-Intimacy Integration

Aftercare is essential. Here are some connection prompts:

- "How are you feeling now—physically, emotionally, spiritually?"
- "Was anything activated or triggered that we should talk about?"
- "What felt most sacred or nourishing about our time together?"
- "How can I honor your energy as we close this connection?"

Close with breath, gratitude, water, or shared silence. Integration is as sacred as penetration.

Appendix C: Unbinding the Sacred Body

A Research-Based Guide to Religious Trauma & Sexual Healing

What is Religious Trauma?

Religious Trauma Syndrome (RTS) is a term used to describe the psychological and emotional harm caused by controlling, fear-based, or punitive religious environments. It often involves:

- **Suppression of sexual identity or expression**
- **Guilt or shame associated with bodily pleasure or desire**
- **Fear of divine punishment for sexual thoughts or behaviors**
- **Loss of autonomy due to rigid doctrine or gender roles**

> *Religious trauma rewires how one sees themselves, their body, and their worth—often confusing control for holiness and guilt for virtue.*

Clinical Research & Findings

1. Purity Culture & Shame

Studies show that women raised in strict religious environments often internalize purity myths, associating their worth with virginity or modesty. This leads to:

- Delayed sexual development
- Inability to communicate needs

- Pain during sex (dyspareunia)
- High rates of sexual anxiety and anorgasmia

Source: Klein, C., Apple, K. J., & Kahn, A. S. (2017). *Sexuality in Purity Culture: Fear, Shame, and Internalized Control*. Journal of Psychology & Theology.

2. Religious Trauma & Mental Health

RTS has been compared to PTSD. Symptoms include:

- Anxiety and depression
- Obsessive thoughts around morality
- Body dysmorphia or detachment
- Emotional repression and guilt loops

Source: Winell, M. (2011). *Religious Trauma Syndrome: It's Time to Recognize It*. The British Journal of Psychiatry.

3. LGBTQIA+ Harm and Sexual Repression

Queer individuals face higher rates of religious trauma and suicidal ideation when forced to suppress identity. This trauma often results in:

- Disassociation during intimacy
- Fear-based sexuality
- Internalized hatred toward desire

Source: Sowe, B. J., Taylor, A. J., & Brown, J. (2017).

Religious trauma among LGBTQ adults. Journal of Homosexuality, 64(4), 510-528.

Common Sexual Wounds Rooted in Religious Conditioning

Wound	Belief/Conditioning	Resulting Challenge
Pleasure = Sin	"Sex is only for reproduction or marriage"	Guilt, repression, avoidance of desire
Female Submission	"Women must serve men sexually"	Inability to express needs or preferences
Body = Shameful	"Cover up to avoid temptation"	Body shame, detachment, disembodiment
Masturbation = Dirty	"Self-pleasure is selfish or sinful"	Suppression of self-exploration and regulation
Desire = Temptation	"You must flee from lust"	Disconnection from sensuality and self-trust

Pathways to Sexual & Spiritual Reclamation

1. Nervous System Regulation

Unlearning trauma requires nervous system safety.

- Practice **somatic therapy**, breathwork, vagus nerve activation, and grounding
- Recognize that arousal and safety can coexist
- Use self-touch to reconnect with your body on your own terms

2. Deconstructing Doctrine

- Examine inherited beliefs and where they originated
- Ask: *Does this belief bring freedom or fear?*

- Read spiritual texts with a trauma-informed lens

3. Sacred Sexual Education

- Learn about anatomy, pleasure, and consent through holistic, body-affirming frameworks
- Reframe sexuality as a **divine expression of life force energy**
- Understand that **pleasure is a birthright**, not a bargaining chip

4. Embodied Spiritual Practices

- Explore sacred sexuality, tantra, and womb healing
- Reconnect with the divine feminine, inner god/ goddess, or energy that honors both eros and spirit
- Create rituals that revere your body as a temple—not a tomb of guilt

Reflective Journal Prompts

- What sexual beliefs did I inherit from religion, and how have they shaped my relationship with my body?
- What moments of guilt or shame around pleasure do I still carry—and where do they live in my body?
- What does *sovereign sexuality* mean to me?
- How can I begin to reclaim sacred intimacy on my own terms?

Recommended Resources

- **Books:**
  - *Pure* by Linda Kay Klein
  - *Shameless* by Nadia Bolz-Weber
  - *Leaving the Fold* by Marlene Winell
  - *Come As You Are* by Emily Nagoski
  - *Sex God* by Rob Bell
- **Practices & Therapies:**
  - Somatic Experiencing
  - Sacred Sexuality Coaching
  - Trauma-informed Spiritual Counseling
  - Deconstruction Groups (evangelical, queer-affirming, womanist, etc.)

Religious trauma does not make you broken.
Reclaiming your body is not rebellion—it is rebirth.
Your sensuality is not a sin—it is your sanctity.

> *You are not here to be punished for your pleasure. You are here to remember that divinity lives in your desire.*

Appendix D: Pelvic Empowerment Guide

Reclaiming Strength, Sensuality & Somatic Safety in the Womb Space

The pelvis is more than a structural anchor—it is the sacred seat of life force, creativity, sensuality, and power. For too long, this space has been ignored, shamed, or only viewed through a lens of reproduction. This guide is designed to reconnect you to your pelvic intelligence and provide scientific, somatic, and spiritual tools for pelvic vitality.

Pelvic Anatomy 101: What You Didn't Learn in Health Class

Structure	Function
Pelvic Floor Muscles	Network of muscles that support bladder, uterus, and bowels. Responsible for continence, sexual pleasure, and core stability.
Vagina	A dynamic, muscular canal that expands with arousal and childbirth. Lined with rugae (folds) for flexibility and elasticity.
Uterus/Womb	Organ of menstruation, pregnancy, and spiritual power. Energetically linked to intuition and creativity.
Cervix	The gateway between womb and vagina. Sensitive nerve endings. Opens during childbirth and ovulation. Plays role in deep orgasm.

Clitoris	Most nerve-dense organ in the body—8,000+ nerve endings. Purely for pleasure, with internal and external structures.

Why Pelvic Health Matters

Pelvic imbalances can affect far more than sexual satisfaction. Common issues include:

- Pain during sex (dyspareunia)
- Urinary incontinence or urgency
- Lower back, hip, or tailbone pain
- Digestive issues like constipation or bloating
- Pelvic organ prolapse
- Numbness, tightness, or disconnection from pleasure

Trauma & the Pelvic Floor

Unprocessed trauma—especially sexual trauma, childbirth trauma, or cultural shame—can manifest in the pelvic bowl through:

- Chronic tightness or hypertonicity
- Guarding or clenching during intimacy
- Disassociation or numbness
- Painful periods or reproductive imbalances
- Emotional shutdown around touch

Clinical Insight:
Studies have shown that trauma survivors often store tension in the pelvis. The pelvic floor may hold "somatic memory" of past violations or fears.
Source: van der Kolk, B. (2014). *The Body Keeps the Score.*

Pelvic Empowerment Practices

1. Pelvic Mapping (Body Literacy)

- Explore the pelvic region gently through self-touch.
- Name and connect to each part: outer labia, inner labia, clitoral hood, vaginal opening, perineum.
- Journal your observations: *What do I feel here? Tension, sensation, emotion?*

2. Breath & Movement

- **Pelvic Breathing**: Inhale into the belly and visualize the breath expanding the pelvic floor. Exhale and allow soft release.
- **Hip Circles & Figure 8s**: Mobilize stuck energy, increase circulation, and restore feminine flow.

3. Yoni Steaming

- Traditional herbal practice used to support menstrual health, clear stagnant energy, and connect with the womb.
- Herbs like mugwort, rosemary, lavender, and calendula support cleansing and calming.

Always consult a provider before steaming, especially if pregnant or postpartum.

4. Kegels—& Reverse Kegels

- Strengthen and relax pelvic floor muscles to support bladder, core, and vaginal tone.
- *Tight ≠ Healthy*—a healthy pelvic floor is both strong and supple.

5. Somatic Womb Connection

- Place one hand on your heart and one on your womb.
- Breathe. Speak love, gratitude, or forgiveness into your body.
- Ask: *Womb, what do you need today?*

The Science of Pelvic Empowerment

- **Improved Blood Flow = Better Arousal**
  Pelvic circulation directly affects clitoral and vaginal engorgement and lubrication. Movement and breath promote sexual vitality.

- **Vagal Nerve & Orgasm**
  The vagus nerve connects the cervix to the brain, bypassing the spinal cord. Cervical stimulation can trigger deep, full-body orgasms through this pathway.

- **Neuroplasticity Through Self-Touch**
  Repeated, mindful touch rewires the brain's sensory map, increasing pleasure pathways and reclaiming agency after trauma.

Source: Porges, S. (2011). Polyvagal Theory.

Pelvic Affirmations

Speak these aloud or write in a journal:

- *My pelvis is the throne of my sacred power.*
- *I release shame and welcome pleasure.*
- *I trust my womb to guide me.*
- *I reclaim every part of me—without apology.*
- *My body is a temple, and I am worthy of worship—starting with myself.*

Sacred Pelvic Self-Care Routine (Weekly)

Day	Practice
Monday	Breathwork + Womb Meditation
Tuesday	Hip Opener Yoga (10 mins)
Wednesday	Journaling: What does my pelvis hold today?
Thursday	Steaming or Abdominal Massage
Friday	Sensual Self-Touch Practice
Saturday	Sacred Dance / Pelvic Mobilization
Sunday	Rest, Nourish, Reflect

Suggested Resources & Tools

- **Books:**

- *Wild Feminine* by Tami Lynn Kent
- *Yoni Shakti* by Uma Dinsmore-Tuli
- *Pelvic Liberation* by Leslie Howard

- **Tools:**
    - Jade or Rose Quartz Yoni Eggs (for energetic + physical toning)
    - Dilators or Pelvic Wands (for rehabilitation or exploration)
    - Guided Pelvic Floor Meditations (via Insight Timer, YouTube, etc.)

Your pelvis is not just a body part. It is an altar. A cauldron. A gateway of creation, death, and rebirth.

To honor your pelvic space is to honor your power.
To reconnect with it is to reclaim what was never truly lost—only waiting for you to come home.

Appendix E: A Man's Guide to Understanding Female Arousal

Honoring Her Biology, Emotions, and Energetic Landscape

Understanding female arousal isn't about mastering performance—it's about deepening presence, attunement, and reverence for the complex interplay of biology, psychology, and energy that shapes a woman's desire. This guide is designed to bridge scientific insights with heart-centered intimacy.

The Science of Female Arousal

1. Female Arousal Is Non-Linear

Unlike the male sexual response, which often follows a more linear pattern, female arousal is circular, dynamic, and often begins long before physical touch. Emotional connection, mental stimulation, and feeling safe are primary.

- **Study Insight**: The Basson Model of Female Sexual Response (Basson, 2000) reveals that desire is often *responsive*, not *spontaneous*, and builds through emotional intimacy and context.

2. Clitoral Stimulation Is Essential

Over 75% of women require clitoral stimulation to reach orgasm. Vaginal penetration alone is rarely enough.

- **Reference**: Studies from Kinsey Institute and research by Dr. Lori Brotto highlight the importance of external stimulation and emotional engagement.

3. The Role of Oxytocin and Trust

Oxytocin, the "bonding hormone," is released through eye contact, cuddling, and loving communication. It creates a sense of safety and emotional closeness that is essential for arousal.

- **Important Note**: If a woman feels emotionally disconnected or unsafe, her body may not physically open or lubricate—even if she's "trying to be in the mood."

Emotional & Energetic Keys

1. Arousal Begins with Emotional Safety
Ask yourself:

- Does she feel emotionally seen?
- Does she feel like her "no" is respected?
- Does she feel like her voice matters?

Safety isn't just about being non-threatening—it's about being *emotionally attuned.*

2. Arousal Thrives with Appreciation
Praise her. Revere her. Speak to her beauty, intellect, spirit, and body—gently, honestly, consistently. The more a woman feels appreciated, the more she blooms.

3. Consent is Ongoing Energy Reading
It's not just "yes" or "no." It's tuning into her body language, her breathing, her tone, and checking in often. Does she feel relaxed? Is her body opening to you? Are her eyes meeting yours with curiosity or withdrawal?

Physical Readiness ≠ Performance Pressure

"If she's not wet, she's not ready."
Moisture is a signal of readiness, but it's not the only one. Arousal can take 15–45 minutes to fully awaken. There's no prize for "getting there" fast. The journey is the gift.

The Power of Foreplay (or, *All* Play)

- Foreplay isn't a prelude—it's a main event.
- Foreplay starts with helping her feel loved in daily life.
- It includes helping with responsibilities, listening deeply, texting her affirmations, or rubbing her back without an agenda.

Words That Touch Her Soul

Use words that help her soften, such as:

- "Take your time."
- "I want to explore you, not rush you."
- "You're safe here."
- "Tell me what feels good."
- "I'm listening to your body as much as your words."

What to Avoid

- **Rushing**: Quickness is often interpreted as disinterest or objectification.
- **Shame**: Never make her feel bad for taking time, needing more, or having boundaries.
- **Comparisons**: Every woman is different. Every moment is different. Stay present.
- **Assumptions**: Ask. Stay curious. Let her guide you with her breath, body, and voice.

Closing Wisdom

A woman's arousal is sacred—it is the language of her nervous system, her emotional safety, and her ancestral womb. To touch her body without touching her soul is to miss the depth of the experience entirely.

The more you listen, the more she opens.
The more you slow down, the deeper she invites you in.
She is not a puzzle to be solved—but a temple to be entered with reverence.

Appendix F: Vaginal Self-Care & Strengthening Routine

Daily Practices to Nourish, Tone, and Empower Your Sacred Center

The vagina and pelvic region are not only central to physical intimacy—they are deeply connected to our emotional, energetic, and spiritual well-being. Regular self-care practices help restore balance, increase pleasure, prevent dysfunction, and reconnect you to your body's sacred wisdom. This guide blends science, tradition, and embodied self-love.

Why Vaginal Self-Care Matters

- **Supports Pelvic Floor Health**: Prevents or reduces issues like incontinence, prolapse, or painful intercourse.

- **Improves Blood Flow**: Enhances arousal, orgasmic potential, and natural lubrication.

- **Clears Energetic Residue**: Releases stored emotional memory, energetic imprints, and trauma.

- **Promotes Self-Connection**: Rebuilds your relationship with your womb and sensuality.

The Science Behind Strengthening

- **Pelvic Floor Muscles (PFM)**: These muscles support the bladder, uterus, and rectum. Weak PFM may result from childbirth, aging, trauma, or lack of use.

- **Benefits of Strengthening**:

    - Increased vaginal tightness and tone

    - Improved orgasm quality and sexual pleasure

- Enhanced posture and core stability
- Prevention of uterine prolapse and bladder leakage

Daily Self-Care Ritual (15–20 minutes)

1. Vaginal Steaming (2–3x/Week)

Best for detox, clearing stagnation, and reconnecting to your womb.
Use yoni steam herbs like mugwort, rosemary, basil, lavender, and calendula. Sit over the steam and allow your pelvic region to absorb warmth and medicine. Set intentions for emotional and physical release.

2. Pelvic Floor Exercises (Daily)

Kegels:

- Inhale deeply, then contract your pelvic floor muscles as if trying to stop urination.
- Hold for 3–5 seconds, then release. Repeat 10–15 times.
- Gradually increase duration and intensity.

Reverse Kegels:

- Inhale and gently push outward (like bearing down).
- This helps balance tightness and encourages release—important for trauma survivors or those with vaginismus.

Yoni Egg Practice *(Optional)*:

- Insert a yoni egg made of jade or rose quartz and perform gentle squeezes.

- Use only when fully aroused or lubricated, and never during active infection, menstruation, or emotional distress.

3. Womb Massage & Castor Oil Packs

- Warm castor oil gently and apply to the lower abdomen.

- Massage in circular motions. Wrap with a towel and heating pad for 20–30 minutes.

- Supports detox, hormone regulation, and circulation.

Natural Lubrication Support

Support your body's moisture with these additions:

- Drink 8–10 glasses of water daily.

- Consume omega-3s (flaxseed, salmon, chia).

- Use organic, plant-based lubricants during intimacy or self-touch (avoid glycerin-based lubes).

Energetic Hygiene

- **Womb Clearing Visualization**: Imagine a glowing light entering your womb and cleansing it of any old energy, tension, or pain.

- **Affirmations for Pelvic Power**:

    - "My womb is sacred."

    - "I release all that no longer serves me."
    - "I am soft, strong, and sovereign."

- **Moon Cycle Mapping**: Track your menstrual cycle and honor the energetic wisdom of each phase (bleed, follicular, ovulation, luteal).

What to Avoid

- Harsh soaps or douching (disrupts vaginal microbiome)
- Excessive tightening practices (may lead to muscle imbalance or pain)
- Ignoring pain—pelvic pain is a sign of imbalance, not a badge of endurance
- Partners who disrespect your boundaries or pressure your body

When to Seek Help

If you're experiencing:

- Persistent dryness or pain
- Discomfort during sex
- Unexplained bleeding
- Vaginal heaviness or prolapse sensations
  Consult a pelvic floor therapist or gynecologist specializing in holistic care.

Weekly Tracker

Day	Steaming	Kegels	Massage	Affirmations	Notes
Monday	☐	☐	☐	☐	
Tuesday	☐	☐	☐	☐	
Wednesday	☐	☐	☐	☐	
Thursday	☐	☐	☐	☐	
Friday	☐	☐	☐	☐	
Saturday	☐	☐	☐	☐	
Sunday	☐	☐	☐	☐	

Your yoni is not just flesh—it is memory, magic, muscle, and mystery. She deserves devotion. This routine is not about perfection, but presence. Your healing begins with how you return to yourself.

Appendix G: The Pain-Free Pleasure Map

Reclaiming Sensuality Through Somatic Awareness and Nervous System Safety

Not all intimacy should hurt—and yet for many women, pain has been normalized. Whether due to trauma, stress, medical conditions, or lack of proper arousal, pleasure can feel out of reach. This *Pain-Free Pleasure Map* is designed to help individuals and couples build body awareness, honor boundaries, and invite healing through intentional, informed touch.

This is a sacred somatic journey—one that begins with curiosity, not performance.

Why This Matters: The Neurosomatic Science of Pain and Pleasure

- **Pain during sex** can result from conditions like vaginismus, vulvodynia, endometriosis, or pelvic floor dysfunction—but also from unresolved trauma and nervous system dysregulation.

- The **brain interprets safety before it allows pleasure**. If the nervous system is in a state of threat (sympathetic fight/flight or dorsal freeze), arousal becomes biologically *impossible.*

- **Pleasure and pain share neural pathways.** The same region of the brain (insula and anterior cingulate cortex) processes both, which is why trauma or tension can override arousal—even during consensual experiences.

Mapping Your Pleasure Zones (Solo or Partnered)

Use this guide to explore with gentleness, intention, and feedback. Keep a journal or use a mirror if needed.

Reminder: Just because a zone is listed doesn't mean it's required to be touched. Consent and feedback are non-negotiable.

Nervous System Prep: Before Touch Begins

- **Grounding Practice**: Try breathwork (4-7-8 breathing), vagal toning (humming or sighing), or progressive muscle relaxation to regulate your nervous system.

- **Arousal vs. Pressure**: Arousal must come naturally, not from expectation. Wait for lubrication and full consent before exploring more sensitive zones.

- **Create Safety**: Dim lights, play soothing music, use soft blankets, and eliminate distractions.

Area	Touch Types to Explore	Sensation (Pain, Neutral, Tingly, Warm, Electric, etc.)	Notes or Feelings
Scalp	Fingertip circles, light pressure		
Ears & Neck	Nibbles, breath, kissing		
Shoulders	Massage, kneading		
Chest/Breasts	Circular motions, flat palm, light squeeze		
Lower Back	Warm compress, soft caress		
Inner Thighs	Feather touch, scratching		
Vulva (External)	Back of hand, tapping, pulsing		

	pressure
Clitoris (External)	Only after warmup. Try rhythmic tapping or indirect touch
Vagina (Internal)	Use clean fingers or wand. Pause if pain arises
Anus	Only when ready. Light touch outside only

Common Triggers to Watch For

Trigger	Possible Cause	What to Do
Sharp Pain in Vaginal Entry	Tension, tight muscles, lack of arousal	Pause. Breathe. Use warm compress. Seek pelvic floor therapy if persistent.
Shame or Discomfort	Religious guilt, trauma memory	Affirm safety. Use grounding words or breath. Seek support if needed.
Numbness	Disconnection from body, freeze response	Explore non-genital touch first. Move slowly. Build interoception over time.

Affirmations for Healing Touch

- "I do not owe my body to anyone. But I offer it to myself with love."

- "It is safe to feel good."
- "I trust my body to guide me toward what it needs."
- "I welcome slow, sacred pleasure."
- "I release shame, and reclaim sensation."

Partner Tips: Holding Space for a Pain-Free Journey

- **Ask before touching.**
- **Check in often.** "How does this feel?" or "Is this okay?" builds trust.
- **Don't take feedback personally.** Redirection = safety, not rejection.
- **Learn her cues.** Hesitation, holding breath, or tensing means slow down.
- **Use lubrication generously.** More is better—especially during healing.

When to Seek Support

Persistent pain during intimacy is **not normal**. Consult a:

- Pelvic floor physical therapist
- Trauma-informed sex therapist
- Functional medicine gynecologist
- Somatic healer or nervous system coach

Sacred Truth

Pleasure should never be a sacrifice. You are not here to endure touch. You are here to receive, feel, and rise into your body's full, unapologetic YES.

Appendix H: Sexual Safety Self-Assessment

A Reflective Tool for Evaluating Emotional, Physical, and Energetic Readiness

Safety isn't just physical—it's emotional, mental, and spiritual. In a world where sexual experiences are often rushed, normalized without proper consent, or disconnected from true intimacy, this self-assessment is designed to bring awareness to the full spectrum of sexual safety.

Use this tool to pause, reflect, and evaluate whether your current or desired sexual experiences are aligned with your highest well-being.

Question	Yes	No	Sometimes	Notes
Do I feel emotionally safe with my partner before, during, and after sex?	☐	☐	☐	
Can I express my desires, needs, or concerns without fear of rejection or shame?	☐	☐	☐	
Do I feel respected when I set boundaries or say no?	☐	☐	☐	
Does my partner validate my emotional experiences or hold	☐	☐	☐	

space for me?				
Have I been able to process past sexual trauma or painful experiences?	☐	☐	☐	
Do I feel safe being vulnerable or emotionally expressive during sex?	☐	☐	☐	

Section 1: Emotional Safety CheckSection 2: Physical & Medical Safety Check

Question	Yes	No	Unsure	Notes
Do I feel physically comfortable during sexual activity?	☐	☐	☐	
Have I and my partner(s) been tested for STIs/STDs?	☐	☐	☐	
Do I feel empowered to ask about my partner's sexual health history?	☐	☐	☐	
Is contraception or fertility awareness being discussed and respected?	☐	☐	☐	
Do I experience	☐	☐	☐	

pain, discomfort, or disassociation during sex?				
Do I have access to regular gynecological or sexual health care?	☐	☐	☐	

Section 3: Nervous System & Somatic Safety

Question	Yes	No	Sometimes	Notes
Can I feel both relaxed and aroused in my body?	☐	☐	☐	
Do I experience pleasure freely, or do I often dissociate or freeze?	☐	☐	☐	
Do I breathe deeply during sex, or do I hold my breath and brace?	☐	☐	☐	
Is my sexual arousal coming from a place of safety—or fear, loneliness, or obligation?	☐	☐	☐	
Do I ever feel emotionally or physically "shut down" afterward?	☐	☐	☐	
Have I developed practices to soothe or regulate my body after	☐	☐	☐	

intimacy?				

Section 4: Energetic Boundaries & Spiritual Readiness

Question	Yes	No	Unsure	Notes
Do I feel energetically drained or revitalized after sex?	☐	☐	☐	
Am I aware of the energy I exchange with a sexual partner?	☐	☐	☐	
Do I consciously choose who I allow into my sacred space (womb, body, aura)?	☐	☐	☐	
Have I cleared or released energy from previous partners I may still feel tied to?	☐	☐	☐	
Does this connection align with my spiritual beliefs or practices?	☐	☐	☐	
Do I feel like I lose or reclaim power through sexual connection?	☐	☐	☐	

Reflection Prompts:

1. **What was most revealing to me during this assessment?**

2. **Where do I feel the most safe in my sexual experiences?**

3. **Where do I feel the most unsafe?**
4. **What needs or boundaries have I been avoiding expressing?**
5. **How can I begin creating more safety—emotionally, physically, and energetically?**

True sexual empowerment starts with awareness and intention. You are not here to perform. You are here to be honored, loved, seen, and cherished—starting with yourself.

You deserve to feel safe, held, and whole in every touch.

Appendix I: Safety Checklist for Sex During Menstruation

Honoring the Womb's Wisdom: A Guide to Informed, Consensual, and Conscious Choices

Menstruation is a powerful time of release, purification, and heightened sensitivity. While period sex can be sacred, bonding, and even relieving for some, it also comes with unique biological, emotional, and spiritual considerations. This checklist is designed to help you and your partner make empowered, respectful choices around intimacy during menstruation.

PHYSICAL HEALTH & MEDICAL SAFETY

Safety Consideration	Yes	No	Unsure	Notes
Have both partners been tested for STIs/STDs recently?	☐	☐	☐	

Are we using protection (e.g., condom) to minimize risk of infection?	☐	☐	☐	
Is the partner aware that a menstruating body is more vulnerable to bacteria or pathogens due to an open cervix?	☐	☐	☐	
Is there open consent to engage in sex during menstruation—not out of pressure or assumption?	☐	☐	☐	
Have we discussed the use of towels, menstrual cups, or positioning to reduce mess or discomfort?	☐	☐	☐	

Did you know? During menstruation, the cervix is slightly dilated, increasing susceptibility to infections, inflammation, and discomfort. Proper hygiene, communication, and gentleness are crucial.

SOMATIC AWARENESS & COMFORT

Self-Check	Yes	No	Sometimes	Notes
Am I currently experiencing cramps, tenderness, or fatigue that would make sex uncomfortable?	☐	☐	☐	
Do I feel energetically open to receiving	☐	☐	☐	

physical intimacy at this time?				
Do I feel grounded, safe, and connected in my body?	☐	☐	☐	
Do I feel like I am honoring my body's needs instead of overriding them for my partner's pleasure?	☐	☐	☐	

EMOTIONAL CONSENT & CONNECTION

Emotional Clarity	Yes	No	Sometimes	Notes
Have I clearly communicated how I feel about period sex?	☐	☐	☐	
Do I feel respected and cherished regardless of my cycle phase?	☐	☐	☐	
Does my partner prioritize my comfort, arousal, and readiness over their own timeline?	☐	☐	☐	
Is this experience bringing us closer—not creating confusion, pressure, or resentment?	☐	☐	☐	

SPIRITUAL & ENERGETIC RESPECT

Energetic Check-In	Yes	No	Unsure	Notes
Do we honor menstruation as a sacred time of release and renewal?	☐	☐	☐	
Am I aware that energy exchange during menstruation can be more potent, sensitive, and imprinting?	☐	☐	☐	
Have I taken time to cleanse or ground after intimacy, emotionally or energetically?	☐	☐	☐	
Is this an act of reverence and healing—not bypassing or denial?	☐	☐	☐	

Womb Wisdom Reminder: Your body is shedding what no longer serves. Only allow energy in that feels like love, care, and aligned reciprocity.

REFLECTION PROMPTS:

- What emotions or beliefs arise for me around period sex?
- Is this something I do to please others, or because I feel empowered and open?
- How can I make intimacy during menstruation more sacred, safe, and soul-nourishing?

AFFIRMATION

"My womb is sacred in all seasons. I honor her flow, her voice, and her boundaries—every day of the month."

Appendix J: Ritual for Reclaiming Menstrual Wisdom

A Womb-Centered Ceremony for Sacred Reconnection, Release & Reclamation

For generations, menstruation has been shamed, silenced, and pathologized. But the blood that flows from the womb is not a curse—it is an offering. This ritual is designed to help you return to the truth of your cycle: that menstruation is a sacred act of release, renewal, and intuitive power. Reclaiming menstrual wisdom allows us to reclaim our body, voice, and sovereignty.

Timing the Ritual

- Ideal during **Day 1 to Day 3** of your menstrual cycle (the heaviest flow)
- Best performed in a **quiet, undisturbed space**
- May be done **alone or in sacred sisterhood** with shared intention

Materials You'll Need

- A **journal or sacred notebook**
- A **red cloth or shawl** to represent your womb
- **A bowl or jar** if you collect your menstrual blood (optional)
- **Candles** (preferably red or white)
- **Herbal tea** (such as raspberry leaf, nettle, mugwort, or rose)

- **Essential oils** (lavender, ylang ylang, or clary sage)
- **Crystals** (carnelian, moonstone, rose quartz, bloodstone)
- A **mirror**, if part of your intention is self-witnessing
- A **music playlist** with soft womb-healing frequencies or drumming

Step-by-Step Ritual

1. Prepare the Space

- Cleanse with smoke (sage, mugwort, or palo santo), sound, or intention
- Create a small altar with your items
- Dress comfortably—consider wearing red or flowing fabrics

Affirmation: *"I call my spirit back to my body. I reclaim all power and wisdom tied to my blood."*

2. Open Sacred Space

- Light your candle
- Breathe deeply into your womb space, placing one hand on your heart and one on your lower belly
- Invite your ancestors, guides, or your higher self into the space
- Call in the four elements: **Earth** (stability), **Water** (flow), **Fire** (transformation), **Air** (wisdom)

3. Womb Connection Meditation

Sit or lie down in stillness. Visualize your womb glowing in deep red or gold. With each breath, imagine this space softening, opening, and releasing. Whisper:

- *"I honor your rhythm."*
- *"I trust your timing."*
- *"You are sacred—not shameful."*

Journal any sensations, emotions, or memories that arise. Allow yourself to cry, laugh, bleed, or just be.

4. Blood Offering (Optional)

If you collect your menstrual blood (using a cup, cloth, or otherwise), you may:

- Offer it to plants or the Earth in a private outdoor space
- Paint with it on paper as a reclamation of ancestral art
- Speak your affirmations over it as a living, sacred elixir

> **Menstrual blood is rich in stem cells, growth factors, and life-force energy. Many indigenous cultures considered it holy.** This practice is not about fetish—but reverence.

5. Write a Womb Letter

Address it to:

- Your current womb
- Your inner girl who first bled

- The ancestral line of women before you
- The daughters to come

Prompt Ideas:

- "What do you want me to remember?"
- "What have I ignored or pushed away about you?"
- "What wisdom do you carry for me?"
- "What are you ready to release?"

6. Movement Integration

Turn on music and move freely. Let your hips sway. Let your breath guide you. Dance for your blood. Dance for your rebirth. Dance for the women who couldn't.

7. Close the Ritual

- Thank your womb and your blood
- Blow out the candle
- Wrap yourself in warmth
- Sip your tea and rest
- Carry this reverence into your next cycle

Reflection Prompts (Post-Ritual)

- What stories about menstruation have I inherited? Are they still mine?

- In what ways have I silenced or punished my womb?
- How can I honor my cycle as sacred moving forward?

Did You Know? (Science Behind Menstruation & Energy Sensitivity)

- During menstruation, **the cervix dilates slightly**, increasing energetic and microbial permeability.
- Hormonal changes during the menstrual phase **heighten intuitive and emotional perception**.
- Studies show that **women may experience shifts in sensory processing** during their cycle (Derntl et al., 2008).
- **Menstrual blood contains mesenchymal stem cells** (Meng et al., 2007), which are regenerative and unique to the womb lining.

Appendix K: Rewriting Your Sexual Beliefs – A Self-Reflection Journal

Unlearning the Shame. Reclaiming the Sacred. Rewriting Your Truth.

Our beliefs about sex are rarely born from inner wisdom—they are inherited. Conditioned. Taught through religion, culture, trauma, silence, and experience. Many of us carry sexual scripts that are rooted in fear, control, shame, or suppression. This journal is a safe space to deconstruct those beliefs and rewrite your own sacred truths.

This process is not about judgment—it's about *liberation.*

Journal Instructions

- Find a quiet, safe, and comfortable space.
- Approach this work with radical honesty and compassion.
- Write without censoring or editing yourself.
- You may revisit and re-answer these prompts across different seasons of healing.

Part 1: Uncovering the Origins

1. **What were the first messages I received about sex? Who did they come from?**

2. **Was sex talked about in my household growing up? If yes, how? If not, how did the silence shape me?**

3. **What did religion, culture, or school teach me**

about sexual "purity"?

4. **Was pleasure discussed as a right or a reward? Was it something to be feared or celebrated?**

5. **How did gender roles impact how I saw myself as a sexual being?**

6. **What did I learn (consciously or unconsciously) about what my body was "supposed" to do or feel?**

Part 2: Exploring Current Beliefs

7. **How do I feel about sex now? Safe? Anxious? Ashamed? Empowered? Numb?**

8. **Do I see sex as a tool, a weapon, a performance, a connection, a chore, or something else?**

9. **What are my current beliefs about masturbation, orgasm, and solo pleasure?**

10. **How do I define "good sex"? Whose definition am I using?**

11. **What makes me feel sexually safe? What makes me feel unsafe?**

12. **Do I feel free to express my needs and desires in partnership—or do I shrink and silence myself? Why?**

Part 3: Rewriting the Narrative

13. **What sexual beliefs am I ready to release? (List them all.)**

14. **What new sexual truths do I want to adopt, embody, and anchor into my life?**

15. **How would my sexual life change if I fully trusted my body, my intuition, and my desires?**

16. **What does *liberated intimacy* look and feel like for me?**

17. **What affirmations can I use to reprogram my sexual mindset? (Examples below)**

Sample Affirmations for Sexual Liberation

- *My pleasure is holy and healing.*
- *I am safe in my body. I am safe in my desire.*
- *I do not need permission to know myself intimately.*
- *Shame is not mine to carry. I release what never belonged to me.*
- *I am allowed to be both sacred and sensual.*
- *Consent is clarity. Connection is choice.*

Closing Ritual

After completing your journal entries, light a candle and speak one new belief aloud.

Place your hands over your womb, heart, or chest and say:

> *"I release the old. I reclaim the sacred. I write a new truth—in my body, in my blood, in my breath."*

Made in the USA
Monee, IL
05 November 2026

67808605R00177